HISTORY

PIERRE-YVES DONZÉ

HISTORY OF THE SWISS WATCH INDUSTRY

FROM JACQUES DAVID TO NICOLAS HAYEK

TRANSLATED BY PIERRE-YVES DONZÉ AND RICHARD WATKINS

PETER LANG
Bern · Berlin · Bruxelles · Frankfurt am Main · New York · Oxford · Wien

Bibliographic information published by die Deutsche Nationalbibliothek
Die Deutsche Nationalbibliothek lists this publication in the Deutsche Nationalbibliografie; detailed bibliographic data is available on the Internet at ‹http://dnb.d-nb.de›.

British Library Cataloguing-in-Publication Data:
A catalogue record for this book is available from The British Library, Great Britain

Library of Congress Cataloging-in-Publication Data: A catalogue record for this book is available.

The publication of this book was kindly supported by TAG Heuer, Branch of LVMH Swiss Manufactures SA and the Gerold und Niklaus Schnitter-Fonds für Technikgeschichte an der ETH Zürich.

Cover Design: Didier Studer, Peter Lang AG

Third edition

ISBN 978-3-0343-1645-3 pb.   ISBN 978-3-0351-0807-1 eBook

This publication has been peer reviewed.

Hochfeldstrasse 32, CH-3012 Bern, Switzerland
info@peterlang.com, www.peterlang.com

Printed in Switzerland

Table of Contents

Introduction

For nearly two centuries, Swiss watches have exerted an insolent domination over the world market. Moreover, despite several crises, this supremacy has never been successfully challenged. The success story of the Swiss watch industry has been and still is largely explained as the result of a long tradition of manufacturing precision instruments, a widely shared technical culture, and an industrial organization as a flexible production system which enabled it to answer all the needs of customers. However, this traditional account, currently kept alive by the marketing strategies of watch companies and highlighting a kind of a timeless "Swiss genius", has to be reconsidered in the light of economic history.

Watchmaking is certainly one of the oldest and most representative industries of Switzerland. A quick glance at the evolution of the foreign trade statistics of the country between 1840 and 2000 makes this importance evident (Table 1). During these two centuries of history, watchmaking is indeed one of the four main Swiss export industries. Together with textiles, machines and chemicals, it largely contributed to making Switzerland one of the richest countries of the world.

The structure of foreign trade shows that watchmaking is, after textiles, the second largest export industry of Switzerland between 1840 and 1937, and even the first in 1953. Moreover, its importance tends to strengthen during these years, with the percentage of exports growing from 8.2% to 21.1%. In the second part of the 20th century, watchmaking is third, below chemicals and machines, two sectors whose growth was particularly high after the war. As for its relative importance, it certainly appears to be decreasing, but this fall-off shows above all a general expansion of the Swiss economy, especially characterized by the dynamism of many sectors, as shown in Table 1 with the sharp increase of "other industries".

The weight of watchmaking in the domestic economy, and the importance watchmaking companies have attached to their own history in their PR policy since the beginning of the 1990s, gave birth to many books and publications. Yet paradoxically, its general history is still unknown and not easily to access. Books are usually limited to a firm, a region or an individual, so that it is difficult to have an overview of the history of the Swiss watch industry in the long run. So, the aim of this book is to offer a general history

Table 1: Relative importance of watchmaking in Swiss foreign trade, 1840–2000 (value as a %)

	1840	1890	1912	1937	1953	1970	2000
Textiles	72.5	57.2	44.1	20.1	16.0	9.2	2.9
Watchmaking	*8.2*	*14.2*	*13.0*	*18.1*	*21.2*	*11.8*	*7.6*
Machines	0.1	3.2	8.1	16.1	20.7	30.4	27.3
Chemicals	0.4	2.3	4.7	15.5	16.3	21.0	26.4
Other	18.8	23.1	30.1	30.2	25.8	27.6	35.9

Source: Veyrassat, Béatrice, "Commerce extérieur", *Dictionnaire historique de la Suisse*, <www.dhs.ch> (site accessed 21 June 2009) and official Swiss foreign trade statistics, <www.bfs .admin.ch> (site accessed 21 June 2009).

of Swiss watchmaking since the middle of the 19th century. The approach adopted here is that of economic and social history. It focuses on the particular structure of this business (industrial districts, *Statut horloger* and groups of firms), as well as on the technical evolution of products (pocket watches, wristwatches and quartz watches), export outlets, rival industries (British, American, then Japanese), the intervention of public authorities (cartels and liberalization) and the relationships with workers' unions.

During the years 1800–2000, the Swiss watch industry faced two main challenges, two revolutions which threatened its existence and led to two major reconversion crises.

The first challenge was industrialization. It occurred during the 1870s and 1880s when the American watchmakers, mainly the Waltham Watch Co. and the Elgin Watch Co., forced Swiss watchmakers to adapt their system of production, with the introduction of machines and the construction of modern factories. The issue was the production of watches: the Swiss had to learn to manufacture standardized products. It was definitely a challenge, which caused considerable ferment amongst Swiss watchmakers until the 1900s. Led by a few industrialists, among whom was Jacques David, technical director of the company Longines, the Swiss watchmakers succeeded in adapting their system of production and overcoming American competition. Around 1900, they controlled about 90% of the world market.

The second challenge was marketing. Seen from a technical point of view, it appears to be what is usually called the "quartz revolution" and led to what is commonly described in Switzerland as the "watch crisis" (*crise horlogère*). However, this challenge went beyond its technical dimension. A new adaptation of the structures of the Swiss watch industry became a necessity, not only due to the launch of quartz watches on world market in the 1970s, but also because powerful industrial groups emerged as competitors, such as Seiko in Japan and Timex in the United States. The issue was not being able to produce watches but being able to sell them. Thus, the 1970s and 1980s are characterized by a restructuring of the Swiss watch industry, to follow the new rules of marketing. New business leaders, among whom was Nicolas G. Hayek, took charge of industrial groups in which they rationalized production and reinvented commercial policy. This marketing revolution enabled Switzerland to strengthen its dominant position on the world market at the beginning of the 21st century.

If one is to really personify the Swiss watch industry between 1800 and 2000, it would be necessary to add a third person: Sydney de Coulon, general director then delegate of the Board of Directors (*administrateur-délégué*) of Ébauches SA from 1932 to 1964. He is the man who personifies the cartel and the *Statut horloger*. He embodies the Swiss watch industry which successfully took on the American challenge of industrialization and then organized itself within rigid structures in order to protect its comparative advantage on the world market. In this context of a real bureaucratization of watchmaking, the domination of Swiss watchmakers on the world market continued until the emergence of competitors in the United States and Japan during the 1960s.

This book is the result of many years spent studying watch history, an activity which led me to consult many archives and to collaborate with numerous persons. It is unfortunately not possible to personally thank all who welcomed me in their institutions, allowed me to interview them and contributed to my work. However, I would like to give special thanks to the whole staff of the Musée international d'horlogerie, at La Chaux-de-Fonds (Switzerland), particularly to Jean-Michel Piguet and Laurence Bodenmann, Patrick Rérat from the University of Neuchâtel, and Patrick Linder, director of the Jura Bernois Chamber of Commerce, for their kind help and availability. Without them, it would not have been possible to finish this book. Also, I sincerely thank Alain Cortat for his remarks, comments and suggestions made on previous versions of the manuscript. Finally, I want to thank Richard Watkins for his kind help translating this book and making it available to a worldwide audience.

This third edition is an updated and slightly corrected version of previous editions. I thank all the readers for their very helpful comments and constructive critics.

Kyoto, October 2014

Map 1: Switzerland and the Watch Industrial District

Source: Designed by Patrick Rérat, University of Neuchâtel, Switzerland

Chapter 1

The Swiss Watch Industry during the first part of the 19th century (1800–1870)

According to tradition, the origin of Swiss watchmaking goes back to its emergence and development in the city of Geneva in the second part of the 16th century.[1] At that time, London, Paris, and the southern part of Germany were the main places in Europe where watches were manufactured. The success of watchmaking in Geneva, and then in all the Jura Mountains, enabled Switzerland to assert itself as the main challenger of Great Britain in the 18th century.

The development of watchmaking in Geneva has a twofold origin. First, the goldsmith's and silversmith's craft must be taken into account, as it was an artisanal activity with an international renown since the Middle Ages. Second, the first wave of protestant immigrants, especially from France (Huguenot refugees) during the 16th century, played a key role, as their arrival gave Geneva new technical know-how, as well as capital and commercial networks which favored the growth of watchmaking in Geneva. Under the influence of Jean Calvin, the passing of luxury laws (*lois somptuaires*) restricting the wearing of jewels (1560), and then the banning of goldsmiths, silversmiths and watchmakers from making religious items such as crosses and chalices (1566) led these craftsmen to redirect their work towards the decoration of watches. The coming together of goldsmiths and watchmakers is at the root of the development of a watch industry in Geneva. Due to this growth, watchmakers soon organized – and protected – their profession, adopting their first guild regulations in 1601. Access to the craft was restricted, as masters could employ only one apprentice, but open to foreigners; indeed, none of the craftsmen who signed the 1601 regulations was a native born Genevan. One must also stress here that the protestant immigration included not only highly skilled artisans and capitalists but also common workers. This made cheap labour available to the domestic proto-industry and contributed to its growth.

The production of watches in Geneva boomed during the 17th and 18th centuries. The annual output is estimated at 5,000 pieces in 1686, rising to 85,000 pieces in 1781, of which 40,000 were in gold and 45,000 in silver cases.[2] This high growth led to some organizational mutations. First, activities were specialized and division of labor spread in watchmaking. For example, craftsmen specializing in the production of springs appeared in the 1660s. Other artisans gathered together into new guilds, as did tool makers (1687), case makers (1698) and engravers (1716). As for watchmakers, who dominated the *Fabrique horlogère*,[3] they tried to maintain their control on that industry. In 1687, they limited access to the mastership, that is, the right to train apprentices, to citizens and bourgeois of Geneva.[4] They also attempted to put an end to the diffusion of watchmaking out of the city into the surrounding countryside, as well as to the arrival of new entrants, especially foreigners. These strategies aimed to save the most profitable and most prestigious part of the work for themselves: the finishing of watches (*finissage*).[5] Lower classes (natives and inhabitants) were only allowed to practice less profitable professions, like the fabrication of parts or the assembling of cases. This class division of labor in watchmaking is at the heart of the social conflicts which troubled Geneva continuously throughout the 18th century. Natives (1738) and then inhabitants (1782) were by turns readmitted to the mastership.

Eventually, the very strong growth of production during the second part of the 18th century caused this system to fail, and production was reorganized to enable an increase of the workforce. The new guild regulations adopted in 1785 opened up some professions to women, and in 1799 more than one thousand were active in this sector.[6] Apprenticeship rules were relaxed and it became possible to train in only one part of the production process. Also, in 1788 the Society of the Arts set up a dial-work (*cadrature*) workshop to train apprentices in a sufficient numbers.[7]

The other main characteristic of the failure of the *Fabrique horlogère* is the appearance of new production centers outside the city of Geneva, in country areas where there were no guilds and where it was possible to freely organize the production of watches. This is precisely the pattern of Le Sentier (Vallée de Joux, in the canton of Vaud), the region of Le Locle and La Chaux-de-Fonds (canton of Neuchâtel) and Saint-Imier (canton of Bern). The diffusion of watchmaking in the Jura Mountains is usually explained as the consequence of its growth at Geneva. In order to benefit from more liberal conditions operating there and in neighboring France, where there were no guilds, watchmakers from Geneva are said to have begun outsourcing work out of their city and so contributed to the diffusion of know-how. However, the emergence of watchmaking in the Jura Mountains is not the only result of a delocalization of the watch industry of Geneva. It also comes from an endogenous development process.

This region enjoyed its own commercial networks, organized on a global scale by some of the trading families of the city of Neuchâtel (de Pury, de Pourtalès, de Coulon, etc.), and these offered ways to sell watches. Knowledge came from outside the region. It was transmitted by watchmakers who stayed there, as for example Simon Fornet, a master of horology from Geneva who settled in Saint-Imier during the 1760s.[8] The wealthiest of these domestic watchmakers also made journeys to Paris or London, where they could practice with the elite watchmakers. For example, Jean-Pierre Droz (1709–1780), born at Renan, in the small valley of Saint-Imier, stayed with watchmakers in Basel and Paris after having finished his apprenticeship at La Chaux-de-Fonds. And Jacques-Frédéric Houriet (1743–1830), son of a watchmaker of Le Locle, went to Paris in 1759, where he pursued his training in the workshop of Julien Le Roy, watchmaker to the King, before coming back to Switzerland, to his hometown Le Locle, where he continued his career.[9] The technology and know-how acquired were then passed on to other local artisans, within family or business relationships.[10] Between 1798 and 1804, 82 Swiss watchmakers, among whom 53 from the Canton of Vaud and 20 from the Canton of Neuchâtel, were granted residence permits to settle at Geneva.[11] Their career is unknown but some of them obviously came back to their home towns after having acquired knowledge. So watch production had a strong growth in the Jura Mountains in the 18th century, an industrial development which was supported both by a favored access to commercial networks and by an active policy of gaining knowledge and skills. According to Frédéric Scheurer, the number of watchmakers in the canton of Neuchâtel went from 464 in 1752 to 3,634 in 1788.[12]

1.1 The triumph of établissage

From when it appeared and spread in the Jura Mountains during the 17th and 18th centuries, until the emergence of the first factories in the last third of the 19th century, the Swiss watch industry was organized in a system called *établissage* (Figure 1).

The key-person of this system was the *établisseur*: while being a middleman between manufacturers and the market, he controlled the operation of the system. He was not an industrialist, but rather a trader. He distributed work to various subcontractors, who themselves sometimes

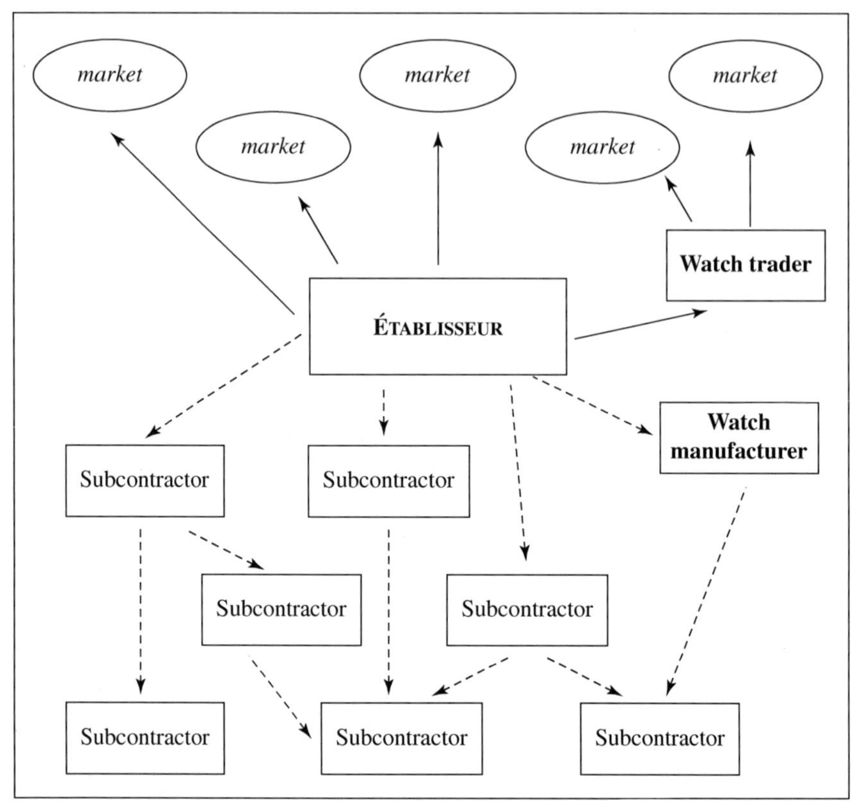

Figure 1: The établissage system of production in the Swiss watch industry
Source: Designed by the author.

outsourced part of the work to second-rank subcontractors; as for example case makers who had polishing and enameling done by others. Then he gathered the parts produced outside (cases, movements, dials, hands, etc.) and assembled them into a finished watch within his own workshop, called a *comptoir d'horlogerie*. Some *établisseurs* also outsourced the assembling to watch manufacturers (*fabricants d'horlogerie*) who sold the finished watches back to the *établisseurs* because they had no access to markets. Thus the production of watches was an extremely fragmented and complex system which was embodied in many and various forms.

The fast growth of this industry during the 19th century, mainly due to the endless expansion of markets, especially in the United States, did not fundamentally challenge this system of production. Of course, a general trend towards a larger division of labor characterizes the second part of the 18th century. However, the increase in demand did not lead to the industrialization of production, but on the contrary to an extension of the specialization of activities which reinforced the system of *établissage*. The manufacture of watches became an activity more and more fragmented between numerous small workshops, sometimes set up within the home of the worker. The number of different operations needed to produce a watch is estimated at 54 around 1830 rising to about one hundred around 1870.[13] An extreme specialization of labor occurred until the 1870s, each home worker focusing on one of the more and more reduced parts of the production process. According to commercial yearbooks, in the city of La Chaux-de-Fonds the number of different occupations in watchmaking rose to 67 in 1870 (see Table 2). Overall, there were then 1,308 autonomous workshops in this town alone. They were essentially small businesses active only in a very limited part of the manufacturing process. The dominance of home work and small workshops is a general characteristic of the Swiss watch industry. For all the country, it is estimated that half of the workers in this sector worked at home in 1870, and 80% of those active in workshops were engaged in firms employing 10 or fewer workers.[14] Even if these various workshops were theoretically autonomous firms, there was a very strong interdependence between them, because they were part of business and outsourcing networks which severely limited their autonomy.

The second function fulfilled by the *établisseur* was the marketing of watches. He was above all a trader with easy access to export outlets. He generally did not work alone but rather closely with family and business relationships. The most typical model is the presence in Switzerland (usually in a watchmaking center like Geneva, La Chaux-de-Fonds or Le Locle) of a family member who supervised the production of watches, and the settlement of other family members elsewhere in the world (Germany, Italy, Great Britain, United States, etc.), where they marketed the products received from Switzerland. Sometimes *établisseurs* dealt with independent watch traders who were intermediaries for accessing specific outlets. Finally, it must be emphasized that the *établisseurs'* families involved in watchmaking in the first part of the 19th century were often connected with each other by matrimonial ties, so that it is possible to describe it as a family "nebula".[15] Moreover, they were often old trading families which had accumulated a huge capital under the Ancien Regime, thanks to the textile trade. This wealth made them the key persons of the *établissage* system of production: they held the capital which was necessary to

Table 2: Watchmaking workshops at La Chaux-de-Fonds, 1870

Activity	Number of workshops
Adjusters	49
Assortments[16]	19
Balance springs	2
Brass smoothing and polishing	15
Broaches and reamers	1
Case dome makers	20
Case dome polishing	17
Case enameling	3
Case engine turning	19
Case engraving and engine turning	50
Case joint fitting	2
Case pendants, rings and crowns	5
Case polishing and finishing	57
Case repairers	4
Case springs	48
Casing movements	9
Centre-wheel caps	7
Cleaners	1
Enamel dial fitting	2
Enamel dial hole making	5
Enamel dial making	33
Enamel dial painters	25
Engravers	45
Escapements	48
Files	1
Finishers and assemblers	202
Finishing	7
Fitting small parts	3
Fusees	2
Gilding	58
Gold case assemblers	30
Hands	19
Jewels and jewel settings	57

(Continued)

Jewel traders	6
Lever escapement rollers	1
Manufacturers and traders	186
Mechanics	9
Metal dial making	10
Metal dial painters	2
Milling cutters	1
Muffles and plates	1
Nickel smoothing and polishing	15
Parts manufacturers	12
Parts traders	7
Plates	2
Polishing small parts (screws, springs, etc.)	22
Polishing squares	13
Regulator indexes	4
Rough movements and finishing	2
Salesmen and brokers	14
Scrap gold and silver recovery	4
Screws	3
Seconds dial fitting	3
Silver and metal case assemblers	13
Springs	16
Steel polishing	30
Studs	1
Turning arbors	2
Watch glasses	5
Watch makers and repairers	26
Watch plates	3
Wheel and pinion teeth cutting	8
Wheel polishing	14
Winding mechanisms	8
TOTAL	1308

Source: Musée international d'horlogerie de La Chaux-de-Fonds (MIH), *Indicateur suisse d'horlogerie* (Davoine Yearbooks), 1870.

distribute work to small subcontracting workshops which were by and large without assets.

An example of an établisseur: the DuBois family of Le Locle

The Dubois family is a very good illustration of the path followed by many *établisseurs* in Switzerland during the 18th and 19th centuries.[17] They were involved in the trading of cloth from at least the end of the 17th century and for this reason had a presence in several European markets. In the middle of the 18th century, a member of the family, Moïse DuBois (1699–1774), added clocks and watches to his business, but he did not produce them as yet (1751). His own son, Philippe DuBois (1738–1808), went into watchmaking for the family outlets and in 1785 he founded a *comptoir* with his sons in the town of Le Locle. The DuBois had business relationships in many European trading centers. For example, two of Philippe's brothers, Guillaume (1726–1794) and Abram (1728–1752), moved to London where they settled and married British women. Their offspring, especially Edouard (1774–1850), son of Guillaume, maintained close relationships with their Swiss cousins from Le Locle. As for Charles (1774–1850), son of Philippe, he opened a *comptoir* at Amsterdam. In the 1810s, the DuBois were present in all the main trading markets in Europe (Germany, Great Britain, Austria, Kingdom of Naples, Prague, etc.).

The descendants of Philippe established matrimonial connections that were profitable for their business. His son Jules-Henri (1779–1837) married Sophie Vuagneux, the daughter of a trader who exported watches to London. In the next generation, the power was concentrated into the hands of two sons of Jules-Henri. The first one, Jules (1805–1872), was involved in a local royalist movement, became disappointed by the democratic revolution which occurred at Neuchâtel in 1848, and emigrated soon after to Frankfurt, a city in which his family had had business partners since at least the 1780s. He and his offspring carried out the commercialization of production. The second son, Louis (1811–1893), took charge of the supervision of production in Switzerland and supplied his brother with watches. He was close to the local elite among artisan watchmakers. His daughter Cécile married Jules Frédéric Jürgensen (1837–1894), a representative of the family of famous Danish marine chronometer and watch manufacturers established at Le Locle at the beginning of the 19th century. The firm DuBois continued its activities in this way during all of the 20th century.

Why was établissage successful?

Etablissage is not a system of production peculiar to watchmaking, nor to Switzerland. Indeed, similar systems of production can be seen in many regions of the world during the 17th–19th centuries, especially in the textile industry (termed the domestic system, proto-industrialization, or *Verlagsystem*). However, in the middle of the 19th century, the Swiss watch industry differed, in its organization, from its main rival, the United States, where the manufacture of watches took the way of mass production within factories. There are several reasons why Switzerland carried on with *établissage* which led to many theories and controversy among historians who tended to focus on some specific aspects. These aspects can be separated into two different arguments, cultural and economic.

The cultural factors which are used to explain the growth of watchmaking in Switzerland as an *établissage* system focus on the role played by Protestantism and the existence of old technical know-how.[18] As well as the traditional link made with the spirit of capitalism, Protestantism favored a high level of education in most of the population, thanks to the reading of the Bible. Thus, a large number of the inhabitants of Jura Mountains would have had a level of education sufficient for involvement in a semi-industrial activity such as watchmaking, so favoring the emergence of many small production units. Moreover, the learning of foreign languages, particularly of German, would have been widespread and would have favored the commercial success of watchmaking. As for the technical knowledge which had existed prior to watchmaking, it was essentially metallurgy and the working of steel, activities which had been carried out for several centuries in the Jura Mountains and which had been transmitted from generation to generation, giving birth to a local technical culture which helped the emergence of watchmaking at the end of the 17th century. The semi-legendary figure Daniel JeanRichard (1665?–1741), considered as the initiator of watchmaking in the canton of Neuchâtel, fits into this context, as he would have became able to repair the watch of an English traveler, and then produce similar products, thanks to the technical know-how he acquired in the blacksmith's workshop of his family.[19] The high level of education and the mastering of an old technical knowledge are then often seen as key elements which helped many persons of the Jura Mountains to begin producing watches in the 18th century and to manage their own independent workshops.

The problem with this cultural explanation is that it does not specifically characterize the Jura Mountains. Of course, wealthy families from the canton of Neuchâtel, the Vallée de Joux and the Vallon of Saint-Imier probably had a high level of education in the 17th and 18th centuries, but this does not explain in itself the emergence of watchmaking. Also, the link made between the old metallurgical know-how and the existence of a technical culture favorable to the manufacture of watches is not specific to the Jura Mountains. In the 17th century, blacksmithing was an activity spread over all Europe and it is not particular to any territory. However, the importance of these factors should not been denied but rather put into perspective. They are necessary but not sufficient conditions.

The explanation of the diffusion of watchmaking in Switzerland as an *établissage* system based on economic factors appears more convincing. Three main elements must be emphasized. First, the absence of guilds encouraged the diffusion of watchmaking out of urban centers. The strict control they imposed until the end of the 18th century, especially on the access to apprenticeships and the division of labor, as was the case in Geneva, favored its expansion into the Jura Mountains. Thus, watchmaking did not develop in the city of Neuchâtel, where economic activities were controlled by local guilds, but in the countryside where watchmaking could be freely set up. Facing a continuous increase in demand, the traders organized the production of clocks and watches there using the *établissage* system.

Second, the lack of capital outside the urban centers such as Geneva or Neuchâtel must be emphasized. The small towns where watchmaking appeared did not have the capital needed for industrializing production. At its beginning, the manufacture of watches was only a side activity among others in families which still went on farming.[20] Even if it was not difficult for many persons to engage in the production of watches or parts, capital was usually lacking for creating an industrialized firm.

Finally, the existence of old commercial networks supported the distribution of production. The Protestant refugees seems to have played a key-role, the "Huguenot International" as it is called by the historian Herbert Lüthy,[21] making possible the development of commercial activities throughout Europe and the Americas. The famous trading families of Geneva and Neuchâtel were organized as global networks and they added clocks and watches to their other activities. Indeed, they did not specialize in the trading of watches but were general traders involved in many businesses (textile, colonial commodities and slaves) and opened outlets where watch dealers established later.

1.2 The technical evolution of products

The invention of the balance spring by Huygens (1674), which allowed better regulation of the movement and a higher precision, revolutionized the use of watches and led to diverse technological innovations. The 18th century is characterized by the adoption of new mechanisms and new escapements which made possible a miniaturization of the watch. One must underline here the innovations made by Jean Antoine Lépine (1720–1814), especially the adoption of independent bridges instead of a frame with two plates, and that of the cylinder escapement. They made it possible to reduce the volume of the watch and make it more attractive, and the so-called Lépine movements are those mainly used by Swiss watchmakers throughout the 19th century. The attitude towards the technological evolution of the watch, which was dominant in Switzerland during the first part of the 19th century, was characterized both by a quest for excellence and the refusal to mechanize production.

An innovation directed to the quality of products

Generally speaking, the innovations made in the Swiss watch industry until the end of the 19th century focused on improving the quality of products, either on a technical level (the mechanism of the watch) or on a design level (the aesthetics of the watch). They did not aim to simplify and standardize the industry around medium-quality calibers which could be produced with machines, as did the Americans. The watch industry in the United States was indeed born during the 1850s as the result of technology transfer from armory and clock factories, two industries at the origin of the so-called *American system of manufacturing*, and based on the principle of the manufacture of standardized parts by machines. By doing so, it became possible to reduce hand work and to obtain very cheap products. In the watch industry, the first which adopted this system of production was the Waltham Watch Co., whose origin goes back to the years 1850–1854.[22]

However, in Switzerland there was a product innovation rather than a process innovation. At the beginning of the 19th century the elite of the Swiss watchmakers deplored the general trend to a decrease in quality of domestic made watches, due to the continuous growth of demand which led some manufacturers to produce without any concern for the quality of their watches. These renowned artisans denounced the division of labor and the growing specialization of professions which could then be observed within établissage, of which the main risk was, according to them, the extinction of traditional local know-how. Through diverse actions, they tried to support improvement in the quality of watches.

The Société d'émulation patriotique of Neuchâtel is a good example of this collective innovation policy.[23] During the first part of the 19th century, it gave prizes to inventors who participated in the technical improvement of watches by innovations in specific parts of the movement, such as wheels (1832), balance springs (1836) and the escapement (1841). The use of machines was not rejected in itself. Thus, the prize given in 1832 rewarded a new process for improving the teeth of wheels with machines. However, the objective in using machines was not to set up an industrialized system of production. Rather, by the improvement of the quality of parts it allowed, the use of machines strengthened the traditional system of production, *établissage*.

The creation of watchmaking schools is another important element of this innovation policy. Since the middle of the 18th century, the high division of labor had led to the widespread practice of very short apprenticeships, each concerned with a small part of the production process.[24] This habit was called into question in the first part of the 19th century by elite watchmakers who thought this trend to divide and limit technical knowledge was dangerous. Fearing the loss of their local technical culture, they promoted the setting up of watchmaking schools to train watchmakers in the whole process of production of the watch. These institutions proposed a three-year apprenticeship based both on theoretical lessons (physics, astronomy, drawing, mathematics, etc.) and on practice in training workshops set up within the school. Their aim was the creation of the next generations of elite watchmakers necessary to manage *comptoirs* and workshops. The first watchmaking school in Switzerland was opened at Geneva in 1824.[25] The abolition of guilds during the French Revolution necessitated the reorganization of the training system in a new institutional form. The Watchmaking School of Geneva followed the approach taken in the 18th century: its aim was to train complete watchmakers for the local workshops, as did the guild of watchmakers until 1798. Other towns followed the example of Geneva, at first at La Chaux-de-Fonds, where a school opened in 1865 after many years of discussion, and then elsewhere in the Jura Mountains: Saint-Imier (1866), Le Locle (1868), Neuchâtel (1871), Bienne (1873) and Fleurier (1875).[26] In other sectors, the apprenticeship developed in the 19th century as a so-called dual system (parallel training both in a company and in a school). However, the high specialization and the small size of enterprises in watchmaking made it difficult to adopt such a system in this industry.

Despite the chronological gap between Geneva and the rest of Switzerland, the main cause of which was the absence of corporations elsewhere in the Jura Mountains, the ideology subtending the creation of these schools is the same: they were set up to promote technical excellence. Nevertheless, their elitist character did not correspond to the real needs of watchmakers, nor to the expectation of the population, as shown by the example of the Watchmaking School of La Chaux-de-Fonds. It was quite successful and the number of apprentices registered grew up from 15 in 1865 to 35 in 1870, but the full three-year apprenticeship was almost never followed. According to a report of the director, among 140 students who attended the school between 1865 and 1874, only 14 (10%) stayed for three years, while 50 studied for less than one year (35.7%), 51 less than two years (36.4%) and 39 less than three years (27.9%).[27] This inadequacy compared with the need of the market led the school to be severely criticized. In a letter sent in December 1874 to the Municipal Council, the administrative board of the school defended its organization and denounced *"the untoward tendency of the present days which tries to emancipate young men as soon as possible. Many parents, because of the expense, and many young people, because they want to become independent quickly, find the classes of the Watchmaking School too long and prefer an incomplete apprenticeship where reasoning is replaced by routine, in order to be self-sufficient sooner."* The board finished the letter declaring that *"the mission of the Watchmaking School is to be a nursery for serious watchmakers and not to follow the trends of this time, as some people seem to want it to."*[28] Thus, the Swiss watchmaking schools did not modernize their training policy before the 1880s.

Finally, the birth of the Observatory of the canton of Neuchâtel (OCN) also appears as the embodiment of this elite policy. Its creation in 1858 was in response to the necessity for watchmakers to have an exact measure of time, based on astronomical observation.[29] This institution had three main functions. The first was the diffusion of the time to other parts of the country by telegraph, especially to the watchmaking cities (La Chaux-de-Fonds, Le Locle and Saint-Imier). This role was important for watchmakers who could then have access to an official referent time in their own town. Second, the OCN established chronometry competitions which aimed to award precision watches. At the beginning of the 20th century, this took on an important commercial dimension: the getting of an award became an advertising argument for several big enterprises, such as Longines and Omega. Third, the OCN checked the rating of watches, a function gradually transferred to watchmaking schools, as successively happened at Bienne (1878), Saint-Imier (1883), La Chaux-de-Fonds (1887) and Le Locle (1901).[30]

This technical armory set up during the first part of the 19th century by elite watchmakers made it possible for Swiss watches to achieve a high level of quality and to obtain a worldwide reputation of excellence. In the international fairs organized in the middle of the century, the Swiss watchmakers showed an insolent excellence to their British and American rivals.

The hard beginnings of mechanization

The reluctance to industrialize was not shared by all watchmakers. Some of them, rare indeed, tried to promote the use of machines for producing watches from the end of the 18th century.[31] However, these attempts were not numerous and often ended in failure. The best known examples are these of Jean-Jacques Jeanneret-Gris, Georges-Auguste Leschot and Pierre-Frédéric Ingold.

Jean-Jacques Jeanneret-Gris (1755–1827) was an engine turner from Le Locle, whose innovations especially impacted on the French watch industry, particularly on the company of Frédéric Japy, at Beaucourt. After having completed an apprenticeship in 1768 with Abraham-Louis Perrelet, at Le Locle, Japy was employed as a worker by Jean-Jacques Jeanneret-Gris before returning to France in the early 1770s and setting up an ébauches workshop. In 1776 he bought the machines and tools of Jeanneret-Gris, who was defeated by *"the inertia of the artisanal world"*[32], as Swiss watchmakers were not interested in using his machines. In addition, Japy ordered ten other machines from Jeanneret-Gris. This machinery was used as the foundation for his industrialized manufacture of ébauches, which became an important source of supply for Swiss watchmakers throughout the 19th century. In Switzerland, ébauche manufactures also appear to have been the few watch companies to use machines. Thus, the Fabrique d'ébauches de Fontainemelon produced ébauches on an industrial level during the first part of the 19th century. Nevertheless, the high diversity of models it had to produce to fulfill the demands of numerous *établisseurs* prevented it from rationalizing its production: between 1825 and 1870, it produced about one thousand different calibers for its customers.[33]

As an inventor and maker of machine tools, Georges Auguste Leschot (1800–1884)[34] worked at Vacheron-Constantin where he created tools which made it possible to mechanize a part of the production.

However, his influence is hard to estimate, as Vacheron-Constantin never really became an industrialized factory.

Finally, the case of Pierre-Frédéric Ingold (1787–1878) should not be forgotten, as his failure in Switzerland embodies the rejection of mechanization prevailing in watchmaking. Trained as a watchmaker, Ingold worked on the development of machine tools for mass producing some parts of the movement which, thanks to standardization and interchangeability, aimed to enable the transition to industrialized production of watches. Everywhere, he faced failure. When the Swiss rejected his innovations, he went at first to Paris (1838), then to Great Britain (1839), where he perfected his prototypes, but he came up against the conservatism of British watchmakers. Moreover, the British Parliament refused to recognize his firm, the British Watch and Clockmakers' Company. He finally left for the United States in 1845, where he stayed until 1852. It is however unclear if his machines were used or not by American watch makers.[35] When he left the United States, he settled in Paris (1852) and then La Chaux-de-Fonds (1858), where he ended his career.[36]

The opposition of a large part of Swiss watchmakers to the introduction of machines into the production process is not only due to the prevailing conservatism in that industry and the strong will to defend technical excellence. We must also consider an economic factor which certainly played a key-role. Within the *établissage* production system, the numerous small, under-capitalized workshops did not have enough resources to invest in the acquisition of machines. The manufacture of watches and parts was largely made within the family, with limited costs: there were almost no real estate outgoings and the labor force was mainly under-paid family members – spouse and children. The introduction of machines would have necessitated both a large financial investment and a reorganization of the workshop to include salaried workers. Thus, the artisanal production system which was *établissage* enabled small makers to keep independent, while it offered cheap labor for *établisseurs* and watch traders.

1.3 The outlets of the Swiss watch industry: the global market

Due to the lack of Swiss foreign statistics before the 1880s, it is difficult to have an exact view of the relative importance of the different outlets where Swiss watches were sold during the first part of the 19th century. Yet, two particularities of watch exports during this period can be emphasized: the importance of the American market, on the one hand, and the extension of outlets to the whole world, on the other hand. The Swiss watch traders who focused on the European market during the 18th century went out to conquer a world which was becoming limitless, thanks to the development of transportation and communication systems, and the opening of countries to trade.

Swiss traders and watchmakers have been present in the United States since the end of the 18th century. There was in particular a large community of traders from Neuchâtel in Philadelphia in the 1790s, then in New York at the beginning of the 19th century. Even if they were mainly bankers and big traders, that is a *"professional aristocracy"*[37], their presence in the New World made it easy to establish and develop commerce with Switzerland which stimulated the settlement of Swiss watch traders in the cities of the Eastern coast. The case of the *comptoir* of Auguste Agassiz, founded in 1832 at Saint-Imier and which became the company Longines, perfectly embodies this importance of the American market. Indeed, watches produced by Agassiz were mainly sent to the United States through his brother-in-law Auguste Mayor, established at New York as a trader in the 1840s.[38] The continuous expansion of the American market led in the 1830s–1850s to a real *"Americanization of Swiss exports"*[39], as Béatrice Veyrassat put it. Even if it is not possible to quantify this growth, these years were characterized by the opening in the United States of branches of many Swiss trading houses. America was not only the main outlet of the Swiss watchmakers in the middle of the 19th century, but remained a steadily expanding market after the Civil War. The value of Swiss watch exports to this country rose from 8.5 million Swiss francs in 1864 to more than 18 million in 1872.[40]

Despite the strong dependence on the American market, the United States was not the only outlet of the Swiss watch industry. There were Swiss watch traders and makers throughout the world. The main outlet during the 18th century, Europe, still had a considerable importance. The *établisseurs* from the Jura Mountains kept business relationships with the main commercial cities of the continent (Amsterdam, Frankfurt, Livorno, London, Paris, etc.). The feature of the European market during that time was its extension to the East, notably through the development of business relations with Russia and Turkey. Russia was not actually a new market, as several watchmakers and jewelers from Geneva, for example the families Duval and Fazy, were present there in the 18th century. Nevertheless, new traders, working especially for watchmakers in Le Locle (Buhré, Gabus, Tissot and Zénith) and distributing watches on a global scale, arrived in this market after the 1810s, contributing to its expansion. As for Turkey, it is a market where watchmakers from Geneva had been established since the 17th century.[41] The Court of Constantinople was traditionally an important customer of Swiss watches which steadily grew during the 19th century. Finally, the main new overseas outlets were Latin America and East Asia. Watch trade to Latin America was similar to the United States, as the export of watches was in line with a more general commercial flow, directed to colonial trade. In the beginning of the 19th century, Swiss traders settled throughout the subcontinent, particularly on the Atlantic side (Mexico, Brazil and Argentina). Their activities were not limited to watch trade but involved colonial business as a whole. The case of the company Auguste Leuba & Cie embodies this trend. It was founded at Rio de Janeiro in 1822 by Auguste Leuba senior (1798–1860, later a member of the Council of State of Neuchâtel), and joined by his son Auguste Leuba junior in 1828. It focused on the import to Brazil of not only watches and clocks, but also of textile and liquors, and the export to Europe of coffee beans, a business in which it specialized in the second part of the 19th century, until it closed down in 1908.[42]

Yet the really new outlet opened to the Swiss watch industry in the 19th century was the Far East. The Chinese market was controlled in the 1820s by the Bovet brothers, originally from the Val-de-Travers. Their enterprise was organized like a multinational, with an international division of labor between the headquarters in London (from 1851), the production of watches in Switzerland and their commercialization in China, from the city of Canton where one of the brothers had been established since 1822. After the Treaty of Nanking which opened China to free trade with the West (1842), several other watch traders arrived in the market, such as the companies Vaucher Frères, Dimier Frères, Juvet and Courvoisier.[43] Finally, the forced opening of Japan to international trade in 1853 brought the globalization of markets to completion. Settled in Japan since the beginning of the 1860s François Perregaux and James Favre-Brandt, watchmakers of Le Locle, launched this new market and favored the import of pieces from their native town (Nardin, Tissot and Zénith). With the opening of Japan, all nations were from then on open to world trade. Swiss watch traderswere everywhere, even if all the outlets did not have the samecommercial importance.

1.4 Rival nations

During the first part of the 19th century, Switzerland established itself as the main watch producer in the world and soon had a real monopoly on the global market, a predominance which was emphasized at the London (1851 and 1862) and Paris (1855 and 1867) World Fairs. At the beginning of the 1870s, Switzerland had an estimated 70% share of the world production of watches. Its main competitors were France (14%), Great Britain (9%) and the United States (5%).[44] The watch industries of these rival nations had different dynamics: while Great Britain and France were respectively in decline and stagnation, the United States appeared to be the main challenger of Switzerland.

The British watch industry, which dominated the sector at the end of the 18th century by the quality of its products, was not able to modernize its system of production and gradually disappeared from the world market. Except for a few attempts to manufacture watches by the industrial mode of production in the 1870s and 1880s (including Ehrhardt at Birmingham, Rotherhams at Coventry, and the Lancashire Watch Co. at Prescott),[45] from the beginning of the 19th century, British watchmakers had been strongly opposed to any modernization of their system of production. They wanted to stay the best in the niche market of luxury manufactured goods which were nearly unique and over expensive. The British Horological Institute, founded in 1858 to defend the British watch industry from American and Swiss makers, who were flooding the domestic market with simple and cheap watches, rejected all proposals for mechanization and acted as the keeper of technical excellence. For example, in the 1880s it refused to allow the teaching of the use of machine tools to apprentices. The British watch industry could not survive within the context of industrialization and free-trade, so by 1914 it had nearly vanished.[46]

In France, the path to modernization was not so tragic and a transition to mechanized production occurred. Paris and Blois were famous watch and clock production centers in the Ancien Regime but, oriented to the manufacture of luxury goods, they lost their importance afterwards. One of the best examples is undoubtedly Louis Leroy & Cie, a company founded at Paris in 1889 whose roots go back to 1785. Renowned worldwide for the quality of its chronometers and complicated watches, this firm did not take the form of an industrial company and declined at the beginning of the 20th century.[47] During the 18th century, the transfer of technology from Switzerland made possible the emergence of the production of cheap watches, as in the case of the Vallée de l'Arve (Cluses) where subcontracting for the makers of Geneva was wide spread. But the case of the city of Besançon must especially be emphasized, as watchmaking appeared there as a new industrial branch thanks to the immigration of Swiss watchmakers encouraged to settle there by the establishment of the Manufacture Française d'Horlogerie by the new Republic (1793). Throughout the 19th century, several watch companies were founded at Besançon by entrepreneurs with strong links with Swiss watchmakers, especially with the Jewish from the neighboring city of La Chaux-de-Fonds, like Joseph Picard, who opened in 1872 a company which became one of the biggest of Besançon in the 1900s under the name of his successor Paul Levy.[48] Finally, one should mention the ébauches manufacture Japy, at Beaucourt, founded in 1770 and which worked for Swiss watchmakers. Despite its real growth during the 19th century, the French watch industry was not competitive on the world market, probably due to the fact of its focus on the domestic and imperial markets. In 1890, the French export of watches to the United States, then the most competitive market in the world, rose to only 78,000 dollars (4.7% of US imports), while Swiss exports amounted to 1.5 million dollars (88.5%).[49]

The new nation which emerged in watchmaking in the middle of the 19th century and soon challenged Swiss supremacy was the United States, where the first industrialized watch plants appeared. Based on the concept of the interchangeability of machine produced parts, the principle of which was transferred from the manufacturers of weapons and clocks, these watchmaking companies were the first in the world to mass produce cheap watches. The two main enterprises were the Waltham Watch Co. and the Elgin Watch Co., respectively founded in 1854 and 1864. Several other less important watchmaking companies were established but none could challenge the quasi monopoly of the two big American makers, which controlled about 80% of the domestic watch production at the end of the 19th century.[50] Their output was indeed phenomenal. At the Waltham Watch Co., growth was particularly sharp in the 1860s, with production rising from 3,000 watches in 1860–1861 to 91,000 in 1872–1873. As for the Elgin Watch Co., it produced 30,000 pieces in 1867–1868.[51] During the Civil War, a real new industry emerged in the United States. Only a small proportion of this industrial production was exported around the world, where it challenged traditional, expensive and luxurious watches, as it was especially the case in the United Kingdom. The overwhelming part of this new production was sold in the

United States, a continuously expanding market. The American market, their main outlet, was the place where Swiss watchmakers faced a new competitor and a new challenge: that of industrialization.

Notes

1 Belinger, Konqui Marianne, "L'horlogerie à Genève", in Cardinal, Catherine (ed.), *L'homme et le temps en Suisse, 1291–1991*, La Chaux-de-Fonds: Institut l'homme et le temps, 1991, pp. 123–129; and Babel, Antony, *Histoire corporative de l'horlogerie, de l'orfèvrerie, et des industries annexes*, Genève: A. Jullien, 1916.

2 Belinger, Konqui Marianne, "L'horlogerie à Genève", in Cardinal, Catherine (ed.), *L'homme et le temps en Suisse, 1291–1991*, La Chaux-de-Fonds: Institut l'homme et le temps, 1991, p. 125.

3 The "watch factory" (*Fabrique horlogère*) was not a factory as we know it. It was also called a "collective factory". This term describes the industry at Geneva, then elsewhere in Switzerland, and emphasizes the fact that it was organized with a strong division of labor between many independent workshops and small firms.

4 In the 18th century, the class structure of Geneva was based on four different groups. At the top, there were "bourgeois", who were actually old families of the city, and the "citizens", who were also considered as traditional Genevan even if they had fewer privileges than bourgeois. These two groups had all the political power and the economic privileges. Below them, there were foreigners who were accepted to live and work in the city ("inhabitants") and their descendants ("natives"). Basically, the opposition between bourgeois and citizens, on the one hand, and natives and inhabitants, on the other hand, characterizes the social relations in Geneva during the 18th century. (See Mottu-Weber, Liliane, "Genève", *Dictionnaire historique de la Suisse (DHS)*, <www.dhs.ch>, site accessed 19 December 2009).

5 Usually a movement would be assembled and checked as soon as all the parts were made (*remontage*). After which the plates and other brass parts were gilded. The movement was then reassembled and checked (*finissage*), after which it was ready to be cased and sold.

6 Belinger, Konqui Marianne, "L'horlogerie à Genève", in Cardinal, Catherine (ed.), *L'homme et le temps en Suisse, 1291–1991*, La Chaux-de-Fonds: Institut l'homme et le temps, 1991, p. 124.

7 Dial-work consists of the mechanisms placed under the dial. In simple watches it is just the motion work, but in complicated watches it includes repeater, calendar and other mechanisms.

8 Marti, Laurence, *A Region in Time. A socio-economic history of the Swiss valley of St. Imier and the surrounding area, 1700–2007*, Saint-Imier: Edition des Longines, 2007, p. 54.

9 *Dictionnaire historique de la Suisse (DHS)*, <www.dhs.ch> (accessed December 2008).

10 Fallet, Estelle and Cortat, Alain, *Apprendre l'horlogerie dans les Montagnes neuchâteloises, 1740–1810*, La Chaux-de-Fonds: Institut l'homme et le temps, 2001.

11 Ozaki, Mayako, "18 seiki kohan junebu-shi no inyumin ni okeru shusshinchi sokugyo kosei no tenkan to renzoku", *Shakai-keizai-shi gaku*, vol. 71 no. 2, 2005, p. 81.

12 Scheurer, Frédéric, *Les crises de l'industrie horlogère dans le canton de Neuchâtel*, La Neuveville, 1914.

13 Fallet-Scheurer, Marius, *Le travail à domicile dans l'horlogerie suisse et ses industries annexes*, Berne : Imp. de l'Union, 1912, p. 300.

14 Koller, Christophe, *"De la lime à la machine". L'industrialisation et l'Etat au pays de l'horlogerie. Contribution à l'histoire économique et sociale d'une région suisse*, Courrendlin : CSE, 2003, p. 167.

15 Barrelet, Jean-Marc, "De la noce au turbin. Famille et développement de l'horlogerie aux XVIIIe et XIXe siècles", *Musée neuchâtelois*, 1994, p. 216.

16 Assortments (*assortiments*) generally means the components that make up the escapement; for example, the escape-wheel, pallets and roller of a lever escapement.

17 Donzé, Pierre-Yves, "Les industriels horlogers du Locle (1850–1920), un cas représentatif de la diversité du patronat de l'Arc jurassien", in Daumas Jean-Claude (ed.), *Les systèmes productifs dans l'Arc jurassien. Acteurs, pratiques et territoires (XIXe-XXe siècles)*, Besançon, 2005, pp. 61–82.

18 Fragomichelakis, Michel, *Culture technique et développement régional. Les savoir-faire dans l'arc jurassien*, Neuchâtel : ISSP, 1994.

19 Marti, Laurence, *L'invention de l'horloger. De l'histoire au mythe de Daniel JeanRichard*, Lausanne : Antipodes, 2003.

20 Scheurer, Hugues, "Paysans-horlogers : mythe ou réalité ?", in Mayaud, Jean-Luc and Henry, Philippe (ed.), *Horlogeries. Le temps de l'histoire*, Besançon : Annales littéraires de l'Université de Besançon, 1996, pp. 45–54.

21 Lüthy, Herbert, *La Banque Protestante en France de la Révocation de l'Edit de Nantes à la Révolution*, 2 volumes, Paris : SEVPEN, 1959–1961.

22 Hoke, Donald R., *Ingenious Yankees. The Rise of the American System of Manufacturers in the Private Sector*, New York: Columbia University Press, 1990.

23 Petitpierre, Alphonse, *Un demi-siècle de l'histoire économique de Neuchâtel, 1791–1848*, Neuchâtel : Librairie Jules Sandoz, 1871.

24 Fallet, Estelle and Cortat, Alain, *Apprendre l'horlogerie dans les Montagnes neuchâteloises, 1740–1810*, La Chaux-de-Fonds : Institut l'homme et le temps, 2001.

25 Jaquet, Eugène, *L'école d'horlogerie de Genève, 1824–1924*, Genève : Editions Atar, 1924.

26 Simonin, Antoine and Fallet, Estelle (ed.), *Dix écoles d'horlogerie suisses : Chefs-d'œuvre de savoir-faire*, Neuchâtel : Editions Simonin, 2010.

27 Guye, Samuel, *Histoire de l'École d'horlogerie de La Chaux-de-Fonds*, La Chaux-de-Fonds, 1965, p. 43.

28 Municipal Archives of La Chaux-de-Fonds, letter of the Board of the Watchmaking School to the Municipal Council, 10 December 1874.

29 *L'Observatoire cantonal neuchâtelois, 1858–1912*, Neuchâtel : DIP, 1912.

30 Pasquier, Hélène, *La "Recherche et Développement" en horlogerie. Acteurs, stratégies et choix technologiques dans l'Arc jurassien suisse (1900–1970)*, University of Neuchâtel , PhD thesis, 2007, p. 72.

31 Veyrassat, Béatrice, “Manufacturing flexibility in nineteenth-century Switzerland: social and institutional foundations of decline and revival in calico-printing and watchmaking”, in Sabel, Charles F. and Zeitlin, Jonathan (ed.), *World of possibilities. Flexibility and Mass Production in Western Industrialization*, New York: Cambridge University Press, 1997, pp. 188–237.

32 Lamard, Pierre, *Histoire d’un capital familial au 19e siècle : le capital Japy (1777–1910)*, Belfort : Société belfortaine d’émulation, 1988, p. 43.

33 Veyrassat, Béatrice, “Manufacturing flexibility in nineteenth-century Switzerland social and institutional foundations of decline and revival in calico-printing and watchmaking”, in Sabel, Charles F. and Zeitlin, Jonathan (ed.), *World of possibilities. Flexibility and Mass Production in Western Industrialization*, New York: Cambridge University Press, 1997, p. 190.

34 Roth, Barbara, “Georges Leschot”, *DHS*, <www.dhs.ch> (accessed 26 June 2009).

35 Watkins, Richard, *Watchmaking and the American System of Manufacturing*, Tasmania: Richard Watkins, 2009, p. 21, <www.watkinsr.id.au/AmSystem.pdf> (accessed 6 October 2010).

36 Veyrassat, Béatrice, “Manufacturing flexibility in nineteenth-century Switzerland: social and institutional foundations of decline and revival in calico-printing and watchmaking”, in Sabel, Charles F. and Zeitlin, Jonathan (ed.), *World of possibilities. Flexibility and Mass Production in Western Industrialization*, New York: Cambridge University Press, 1997, pp. 190–191.

37 Veyrassat, Béatrice, *Réseaux d’affaires internationaux, émigrations et exportations en Amérique latine au XIXe siècle. Le commerce suisse aux Amériques*, Genève : Droz, 1994, p. 63.

38 Francillon, André, *History of Longines preceded by an essay on the Agassiz comptoir*, Australia: Richard Watkins, available from <www.watkinsr.id.au>, p. 9. (Translation of *Histoire de la fabrique des Longines, précédée d’un essai sur le comptoir Agassiz*, Saint-Imier : Longines, 1947).

39 Veyrassat, Béatrice, *Réseaux d’affaires internationaux, émigrations et exportations en Amérique latine au XIXe siècle. Le commerce suisse aux Amériques*, Genève : Droz, 1994, p. 106.

40 Koller, Christophe, *“De la lime à la machine”. L’industrialisation et l’Etat au pays de l’horlogerie. Contribution à l’histoire économique et sociale d’une région suisse*, Courrendlin : CSE, 2003, p. 114.

41 Kurz, Otto, *European Clocks and Watches in the Near East*, London: Warburg Institute, 1975.

42 Veyrassat, Béatrice, *Réseaux d’affaires internationaux, émigrations et exportations en Amérique latine au XIXe siècle. Le commerce suisse aux Amériques*, Genève : Droz, 1994, pp. 416–417.

43 Chapuis, Alfred, *La montre chinoise*, Neuchâtel : Attinger Frères, 1919.

44 Koller, Christophe, *“De la lime à la machine”. L’industrialisation et l’Etat au pays de l’horlogerie. Contribution à l’histoire économique et sociale d’une région suisse*, Courrendlin : CSE, 2003, p. 103.

45 Cutmore, M, *Watches 1850–1980*, London: David & Charles, 1989, p. 84.

46 Davies, Alun C., “British Watchmaking and the American System”, *Business History*, 1993, n°35/1, pp. 40–54.

47 Exposition internationale de Saint Louis (U.S.A) 1904. Section française. Rapport du Groupe 32, Paris : Comité français des expositions à l'étranger, 1906, pp. 15–16.

48 Exposition internationale de Saint Louis (U.S.A) 1904. Section française. Rapport du Groupe 32, Paris : Comité français des expositions à l'étranger, 1906, p. 17.

49 United States, Bureau of Census, *Foreign Commerce and Navigation of the United States*, Washington: U.S. Govt Print. Off., 1890.

50 Harrold, Michael C., *American Watchmaking. A Technical History of the American Watch Industry, 1850–1930*, Columbia: NAWCC, 1984.

51 Koller, Christophe, *"De la lime à la machine". L'industrialisation et l'Etat au pays de l'horlogerie. Contribution à l'histoire économique et sociale d'une région suisse*, Courrendlin : CSE, 2003, p. 105.

Chapter 2
The challenge of industrialization (1870–1918)

The *établissage* system of production peaked at the beginning of the 1870s. From then on, the Swiss watch industry entered a period of profound modernization of its structures, which is characterized by the assertiveness of the factory and the industrialized mode of production. Mechanization of work and concentration of workers into plants did not however appear as a sudden change. On the contrary, there was a slow but irreversible trend which lasted until the 1910s.

The industrialization of the Swiss watch industry was driven by two external factors: American competition, on the one hand, and the world economic crisis of the 1870s and 1880s, on the other hand. The problem of the American challenge was indeed coupled with the effects of the Great Depression of 1873–1896. This period of crisis created a context which helped the acceptance of the mutation of the production mode in the watch industry. It reinforced the necessity to modernize the manufacture of watches with the introduction of machines as a mean to get cheaper products, and to diversify the outlets in order to reduce the dependence on the American market. Nevertheless, concentration into large organizations was limited. The Swiss watch industry did not give up its existing structure, typical of what economists call the *industrial district*.[52] This concept describes industries concentrated in a region and whose internal organization is based on numerous small and medium sized companies, which are competitors even if interdependent. These industrial systems, such as the textile and shoe industries in Italy, and Silicon Valley in California, are also characterized by a shared technical culture which favors the mobility of workers and the creation of new enterprises. Thus, industrial districts made it possible to manufacture specialized products in a flexible way and to compete with standardized mass production of big companies. In the case of Swiss watchmaking, the industrial district organization enabled watchmakers to dispose of an excessive supply due to a wide range of products on world markets and to hold a quasi monopoly situation around 1900.

The second characteristic of this period, linked to the modernization of the production mode, is the emergence of new collective organizations.

The challenges faced by watchmakers led to the creation and rapid development of associations and groupings which took charge of the defense of the interests of the members in the political field and tried to coordinate the activities of the enterprises in the industrial field. As for workers, they gathered into several associations and trade unions which merged into large organizations at the beginning of the 20th century. The years 1870–1918 see a transition from liberal capitalism, which characterized the *établissage* system of production, to organized capitalism.

2.1 The shock of Philadelphia: the American competitors

The growth of the American watch industry which began in the 1860s became really pronounced in the 1870s. It challenged not only the export of Swiss watches to the United States, its main outlet, but also the worldwide dominant position of Swiss watchmakers. Thus, the American competitors forced the transformation of the production system in Switzerland. American watch companies established themselves on the world market thanks to the industrial system of production. Production was extremely concentrated, as both the Waltham Watch Co. and the Elgin Watch Co. made up the overwhelming part of this industry. It was moreover based on the mass production of cheap watches. At the Waltham Watch Co., output rose from 91,000 items in 1872–1873 to 882,000 in 1889–1890, while it reached 100,000 items in 1879–1880 and 500,000 in 1889–1890 at the Elgin Watch Co.[53] As a comparison, the output of Longines, which was then one of the most modern Swiss watchmakers, was only 20,000 watches in 1885.[54]

This growth had a direct effect on Swiss watch exports to the United States: after having peaked at 18 million francs in 1872, they fell to about 12 million in 1874 and less than 4 million in 1877.[55] This dramatic decrease made it necessary for Swiss watchmakers to react. The Centennial International Exhibition at Philadelphia, held in 1876, gave the opportunity to American industrialists – among whom were watchmakers – to display the dynamism and the modernity of their industry, and for Swiss watchmakers to become aware of that.[56] Among the many leading figures who went across the ocean, were the two delegates sent by watchmakers and the cantons of the Jura Mountains, with the mission to bring back detailed data on American watch factories. They were Jacques David, chief-engineer at Longines, Saint-Imier, and Théodore Gribi, from the company Borel & Courvoisier, Neuchâtel. The organization of this joint delegation to Philadelphia was one of the reasons for the creation of the Société intercantonale des industries du Jura (Intercantonal Society for the Jura Mountains Industries, SIIJ), which in 1876 brought together representatives of the cantons where watchmaking took place. In June 1876, Ernest Francillon, head of Longines, talked of the *"absolute necessity"* to take the opportunity of this exhibition *"to carry out a serious and detailed survey of the organization, the equipment, the financial circumstances and in general of all what relates to American watch companies."*[57]

> Letter of Théodore Gribi, July 1876
>
> "I have examined these last days, as an expert for the Jury, the products and tools of the Waltham watchmaking company and I felt admiration, I must confess, while observing both the different kinds and quality of watches, or the wonderful machines and tools this company showed. It must be admitted that we have let our competitors of the New World outstrip us in many respects, and all Swiss watchmakers who will come here to make enquiries on that point, without any prejudice, will be soon convinced [...]."
>
> Source: David, Jacques, *American and Swiss Watchmaking in 1876*, Tasmania: Richard Watkins, 2003, p. 7.

David and Gribi were fascinated by what they observed in the United States and came back from Philadelphia with a vigorous determination to transform and modernize the Swiss watch industry. Jacques David brought back with him some American watches which he made available to watchmakers and watchmaking schools in order to let them examine American mass produced watches. Above all, he is famous even now for the much talked-about report he wrote together with Gribi.[58] Having visited the three main American watch plants, Waltham Watch (900 workers), Elgin Watch (650 workers) and Springfield Watch (300 workers), David showed the existence of extremely modern manufactures, financed by capitalists and run by salaried managers, where machines were omnipresent. He emphasized that mechanization and interchangeability of parts made it possible for Americans to mass produce cheap watches designed for general customers. In a long technical part, he describes the equipment and manufacturing methods in use in these companies, the principle of which can basically be summed up as *"doing by machine all that can be done so."*[59] The conclusion of the report is a plea for the modernization of Swiss watchmaking through the general use of machines, the standardization of watch calibers, the employment of American mechanics and the development of watchmaking schools. He finally wrote that *"these new factories must be founded in Switzerland and if they are not created here, they will be in the United States and there won't be anything left for us after few years, as Americans already send their watches to our outlets, in Russia, in England, in South America, in Australia and in Japan."*[60] Jules Borel, the employer of Gribi, went further and proposed to the SIIJ in October 1876 to engage an American mechanic having worked for some twenty years at the Waltham Watch Co.: *"This person would come to Switzerland and would be at the disposal of watchmakers and machine-tool makers to lead the reorganization of their production system [...]."*[61] Nevertheless, the committee of the SIIJ decided to wait. The David and Gribi report caused a shock within the Swiss watchmaking business. Many makers continued to defend the artisanal production system and firmly opposed the spread of mechanization. Attached to an idealized image of the watchmaker's work, based on the independence of the home worker and the excellence of handmade work, they rejected industrialization of their industry and preferred to support it through the reinforcement of the quality of products. This negative reaction to the report led the SIIJ committee to decide not to publish it. In addition, a second report written by David mentioned that *"it was agreed that the discussion would not be published in journals, so as not to wake the indiscreet attention of foreign manufacturers and not give them weapons to use against us by too openly making the truth known to them."*[62] So the report may have been suppressed to prevent the Americans from learning about it and taking action against the Swiss.

Even if the SIIJ continued to play a key role in the collective defense of watchmakers' interests, especially on the political field, the modernization of production essentially took place at the level of the individual firm.

Jules F.U. Jürgensen and the belief in technical excellence (1880)

"I believe in the heredity of the hand, in the transmission from fathers to their sons of delicacy of touch, of intimate know-how, of the artistic handling of file and graver. I believe in it for I have noticed thousands of times that fineness in work is, so to speak, patrimonial in many watchmakers' families. I believe in it because even if, on the one hand, I became convinced of the steadiness and the infallibility of the taste of the Parisian worker, a faculty that he raises to a higher personal level by work – but he inherited rather than acquired it and he is given it when he is born, in his cradle he nourishes it with the air he breathes and the objects which he first sees – on the other hand, I am certain it is not possible to create a watchmaking environment, filled with outstanding mechanics, even with the most advanced machines and the best tools. In a nutshell, there is in the manual touch of our workers a certain combination of delicate boldness, of innate grace, of artistic value – which springs from the depths of the soul."

Source: *Catalogue officiel illustré et explicatif de l'Exposition nationale d'horlogerie et internationale de machines et outils employés en horlogerie en juillet 1881 à La Chaux-de-Fonds sous le patronage de la Société d'émulation industrielle*, La Chaux-de-Fonds : Imp. du National suisse, 1881, p. XLIII.

2.2 The structural modernization of Swiss watchmaking

Despite the reticence of some, particularly in the old watchmaking areas such as La Chaux-de-Fonds and Le Locle, the Swiss watch industry experienced a significant mutation of its structures during the years 1870–1918 which was characterized by the shift from the *établissage* system to factory work. Mechanization produced an important increase in productivity and contributed to the strengthening of the Swiss watch industry's competitiveness on the world market: the average number of items produced each year by workers in this sector rose from 65 in 1876 to 170 in 1900 and 330 in 1913.[63]

However, it was only a partial transformation. Big factories as they existed in the United States, and later in Japan, did not appear in Switzerland. On the contrary, the structural modernization of Swiss watchmaking is characterized by the emergence of a dual model, bringing together modern factories and traditional workshops. The hundreds of firms were legally autonomous but strongly interdependent due to their subcontracting relationships. They were organized as an industrial district, a model which allowed the Swiss watch industry to exercise a quasi monopoly on the world market around 1900, thanks to the plethora of different designs made possible by this flexible production mode.

The emergence of the factory

The years 1880–1920 correspond to the assertion of a new production mode characterized by the concentration of workers and mechanization of work. However, instead of the appearance of a few big companies, the growth was in small and medium sized firms. After the passing of the federal law on factories (1877), the central government gained control over these companies. Its main concern was the control of workforce and working conditions (a ban on the employment of children younger than 14 years, daily working time limited to 11 hours, etc.). It also led to the compulsory adoption of factory rules, setting the order within the workshop (working hours, days off, discipline measures such as the ban on singing or smoking while working, etc.) and contributing to the establishment of the power of the boss.[64] However, the kinds of enterprises subjected to the law were not strictly specified, the 1877 law defining a factory as *"any industrial organization where a more or less significant number of workers are simultaneously and regularly employed, outside of their homes or in a closed place."* The meaning of *factory* was more precisely defined in 1891 as *"organizations of more than five workers, using mechanical motors or employing persons younger than 18 years old or presenting some danger to workers' health or life; all the organizations employing more than 10 workers, even if none of the above conditions are fulfilled."*[65] Thus, until 1891 the scope was very large but federal inspectors rarely included watchmaking workshops in their surveys. Statistics published in 1882 and 1888 must then be considered as minimal numbers. Yet a clear increasing trend can be observed which continued until the interwar period.

Table 3: Number of watch factories, 1882–1911

	1882	1888	1895	1901	1911
Factories	72	191	465	647	853

Source: Koller, Christophe, "*De la lime à la machine*". *L'industrialisation et l'Etat au pays de l'horlogerie. Contribution à l'histoire économique et sociale d'une région suisse*, Courrendlin: CSE, 2003, p. 179.

Table 4: Main Swiss watchmaking companies by number of workers, 1905

Enterprise	Sector	Workers
Langendorf SA	Ébauches	1098
Compagnie des Montres Longines	Watches	853
Omega SA	Watches	724
Tavannes Watch Co SA	Watches	609
Zénith SA	Watches	574
Fontainemelon SA	Ébauches	558
Obrecht & Cia	Ébauches	541

Source: Fallet-Scheurer, Marius, *Le travail à domicile dans l'horlogerie suisse et ses industries annexes*, Berne: Imp. de l'Union, 1912, p. 314.

However, these factories were generally speaking small firms. At the beginning of the 20th century, factory work, characterized by the concentration of workers and mechanization, was quite uncommon in the Swiss watch industry. In 1901, the average number of workers per factory reached only 37.5 persons.[66] The production system was then much closer to the workshop model than the factory one. The federal companies survey of 1905 makes it possible to see the small extent of industrialization. There were only seven enterprises employing more than 500 workers in their factories (Table 4) and the biggest one, the Langendorf ébauches factory, had only 1,098 workers. Moreover, among these seven firms, there were only four producers of finished watches, while three were movement makers. Also, the relative importance of these four watchmakers in the overall national watch production was very small. For example, the production of Longines, estimated at 130,000 items in 1905, represented only 1.4% of the Swiss watch export volume in that year, even if it was the second biggest company in the country.[67]

Except for these seven firms which were the biggest in the country, industrialization also occurred within many ébauche and watch factories. In 1901, both of these sectors were indeed the most concentrated: they employed respectively an average number of 68 and 67 workers.[68] These companies, which normally employed cheap labor without qualifications, usually appeared outside the traditional watchmaking areas, mainly at the foot of the Jura Mountains (Biel, Grenchen, Solothurn and Waldenburg) and elsewhere in the canton of Bern. By giving birth to new watchmaking regions, industrialization was accompanied by a territorial restructuring of the watchmaking industrial district. The canton of Neuchâtel, which held the dominant position during the *établissage* era, gave way to the next canton, Bern. Having some of the biggest watch companies in Switzerland, Bern had a growing importance at the end of the 19th century: its share of employment in watchmaking rose from 36.1% in 1870 to 44.4% in 1890 before decreasing to 40.4% in 1900 due to the emergence of new areas (Solothurn and Grenchen).[69] The cities of Biel and Grenchen particularly had a phenomenal growth thanks to the development of watchmaking. At Biel, where the population increased from 5,900 inhabitants in 1850 to 30,100 in 1900, tens of new enterprises opened, among which were the watch companies La Champagne (1854), Heuer (1860), Recta (1897), Aegler – which later became part of Rolex (1878), Louis Brandt & Frère (1880), Concord (1908), Glycine (1914) and Mido (1918); and also the case makers La Centrale (1896) and Maeder-Leschot (1913). As for Grenchen, where the population grew from 1,600 inhabitants in 1850 to 5,200 in 1900, it was the site of the ébauche factories Eterna (1856), A. Schild SA (1896) and A. Michel SA (1898), three companies which became the core of the Ébauches SA trust in the interwar period. Moreover, the multiplication of small firms also occurred in some areas of the Jura Mountains where watchmaking had been rare until then (Delémont, Tavannes and Porrentruy). The presence of the three biggest Swiss watch companies, namely Omega, Longines and Tavannes, embodies the overwhelming importance of the canton of Bern at the beginning of the 20th century.

At Biel, the company Louis Brandt & Frère (Omega) was the main promoter of industrialization.[70] Founded in 1848 at La Chaux-de-Fonds as a *comptoir* by Louis Brandt, it was moved to Biel in 1880 by his sons, Louis-Paul and Charles-César Brandt, with the aim of setting up a modern factory based on the American model, relying on the general use of machine tools and concentrating workers in a factory. In 1890, their enterprise employed 600 workers, within the factory or at home, and produced some 100,000 watches. It had an organization founded on two divisions, production and sales, each directed by one of the brothers. This was the common organization of Swiss watchmaking familial firms during the years 1880–1930. Louis Brandt & Frère faced a dramatic expansion, based on the rationalization of production and the launch of the Omega caliber, developed in 1894 and which came to embody the popularity of the firm. The growth went on in the beginning of the 20th century, with the number of workers rising from 537 in 1900 to 981 in 1914.[71] The family Brandt especially employed women (52% of workers in 1900 and 54% in 1914). The combination of a female workforce and mechanization made it possible to reach an optimal cost of production, as can be testified in the following note written by Louis-Paul Brandt around 1901–1902: *"I ordered from Lambert at Grenchen a machine for turning parts. This operation will be done with two rotary cutters, and we will employ a woman."*[72] As a way to support its growth, Omega reorganized its production in the second part of the 1890s. Together with the metalworking firm Edouard Boillat & Cie, from the nearby village of Reconvilier, they set up an ébauches factory, La Générale (1895), as well as a watch case factory (1896), in order to provide Omega with parts. After the death of the Brandt brothers, the firm was carried on by the third generation under the form of a stockholding company (1903) and it continued to expand: Omega sold 210,000 watch movements in 1910 and 260,000 in 1918, which was about twice its main competitor, Longines.[73]

The company Longines, from Saint-Imier, followed a similar path of development.[74] It came from a small *comptoir* founded in 1832 which was purchased twenty years later by Ernest Francillon, who decided in the 1860s to rationalize production. In 1867, he built a factory and engaged the engineer Jacques David to assist him in the organization of this new manufacture. Little by little, during the 1870s–1900s, the management of Longines concentrated workers within the factory and introduced machines. The number of employees increased from 170 in 1870 to 667 in 1901. As for the production, it was also increasing: the annual output was 20,000 watches in 1885 and 93,000 in 1901.

Finally, Tavannes Watch Co., although later, had a process of development analogous to Omega and Longines. It was born from the desire of the local authorities of Tavannes to develop their small rural village through the introduction of watchmaking. So in 1890 they built a factory which became a new enterprise, Tavannes Watch Co., financed by the watch traders Schwob, from La Chaux-de-Fonds, and managed by Henri Sandoz. The *Journal suisse d'horlogerie,* in an article published in 1913 singing the praises of the company, noted that Sandoz *"ordered from America the huge tools indispensable to this work, lathes, milling machines, etc., which led him to set up a model mechanics workshop where the small specialized machines for watchmaking would be designed and made. He showed them at Geneva, in the machines' hall, during the National Fair of 1896, in front of fascinated visitors."*[75] Based on the American model, this factory had a dazzling growth at the beginning of the 20th century: between 1900 and 1914, the number of workers went from 350 to 1,200, while the daily output rose from 1,000 to 3,200 pieces.

The neighboring canton of Neuchâtel held the fourth biggest watch factory, which employed 500 workers in 1905: the company Zénith.[76]

Founded as an ébauches workshop in 1856 by Georges Favre-Jacot, the factory in Le Locle had a sharp growth due to the early introduction of mechanization and an active innovation policy. A new factory was built in 1881 and enlarged in 1883–1884. During the years 1888–1920, the firm registered 93 patents.[77] Favre-Jacot transformed his company into an industrialized firm during the 1890s and the 1900s, with the aim of increasing output by the systematic use of machines. In 1898 he engaged a former watchmaker, named Breguet, as an industrial statistician in order to control production costs. Also, in 1904 he sent his nephew and son-in-law James Favre to the United States for a study tour. After he returned home, he took charge of the rationalization of production and, above all, adopted a new commercial policy based on advertising (1907). However, this modernization of Zénith had a cost: that of independence. The growth made necessary the creation of a joint-stock company in 1896, with a capital of 1.25 million francs, in which the Banque cantonale neuchâteloise, the most important local bank, had a share. It was then transformed into a public stockholding company with a capital of 2.1 million francs in 1911.

These four biggest watchmaking firms of the 1900s, Omega, Tavannes, Zenith and Longines, should not, however, overshadow the fact that industrialization also occurred within other companies and other areas of the country, even if the process is less emphasized and the scale smaller. The whole of the Jura Mountains were profoundly marked by the transformation of production modes and the emergence of factories. In the Valley of Joux, it was predominantly the company of the LeCoultre brothers, specializing in the manufacture of watch movements, which shifted to mechanization as early as the 1870s. The modernization of this firm had above all the aim to reduce production costs and so to market cheaper watches, as explained Elie LeCoultre: "*Sale prices were always decreasing, both for simple and complicated watches, and this led us to raise output to overcome this, which was possible with new workshops.*"[78] A new factory was built in 1888 and made it possible to engage more employees. The workshop reached 251 persons in 1890 and 314 in 1914. The firm became a stockholding company in 1899 but still remained controlled by the family. Its turnover exceeded one million francs for the first time in 1907.

In the canton of Neuchâtel, usually seen as a canton reluctant to industrialize, one must stress the existence of the ébauche factory of Fontainemelon. Founded in 1793 under the name of the Benguerel & Humbert watch company, it was taken over by the Robert family in 1825 and rapidly expanded after that. In 1913, its output was estimated at one million ébauches, with a workforce of 1,030 employees.[79] Two other companies in the canton of Neuchâtel employed more than one hundred workers in the beginning of the 1890s. Both were ébauche factories: the firms Hahn & Cie, in Le Landeron (134 workers in 1891) and the Manufacture de Chézard (143 workers in 1894).[80]

Similarly, the International Watch Co. (IWC) was founded in 1869 at Schaffhausen, on the Rhine River about 50 km to the North of Zurich. It was and is still the biggest watch company set up outside of the Jura Mountains.[81] The roots of this enterprise go back to an industrial project set up by an American engineer, Florentine A. Jones. Leaving his position at Howard & Co. in 1868, a watch company situated in Massachusetts which was taken over by the Keystone Watchcase Co. in 1903, Jones came to Switzerland with the aim of launching into the production of watches on a large scale, taking advantage of both American technology and Swiss qualified and cheap labor.[82] Nevertheless, he met with resistance from artisan watchmakers in the Jura Mountains and settled at Schaffhausen where he launched his business with a local industrialist, Heinrich Moser, who had owned a watch business at Le Locle since 1829 and engaged in several industrial projects at Schaffhausen. However, this enterprise faced financial difficulties and was taken over by Johannes Rauschenbach in 1880, before experiencing high growth: it employed 104 persons in 1889 and 190 in 1901.

Finally, the trend to industrialization can be observed at La Chaux-de-Fonds, even if many makers were opposed to mechanization, especially among the small subcontractors.[83] Many of the old *établisseur* families, who controlled the production of watches at La Chaux-de-Fonds in the middle of the 19th century, did not shift to mechanization. They gave way to new industrialists who modernized the local industry. For example, the Courvoisier, Othenin-Girard and Gallet families, linked with each other through a dense matrimonial network, were at the heart of the établissage system of production until the 1880s. Little by little, after many years of financial difficulties, they vanished from the watch business at the beginning of the 20th century: Courvoisier Frères closed up in 1931, Girard-Perregaux SA was bought up by the Graef family in 1930 and the company Electa SA (the Gallet family) closed in 1928. The failure of the *établisseurs* of La Chaux-de-Fonds comes from their reluctance to industrialize production. La Chaux-de-Fonds was a real "collective factory" (*fabrique collective*): the commercial directories list 1,299 makers in 1880 and 2,697 in 1900.[84]

They were essentially small workshops specializing in very specific parts of the watch, and they heavily depended on watch makers and traders who dominated the local industry and society. Thus, the shift to industrialization was perceived as a danger which could destabilize the social order and so they refused. Moreover, this economic conservatism was shared by numerous subcontractors, usually without any wealth, for whom the workshop was the only family asset. For them, industrialization meant the end of their autonomy and of their professional activities. This local reluctance to industrialize led some watchmakers to leave La Chaux-de-Fonds and set up new plants elsewhere, as did for example the Brandt brothers, who moved the family *comptoir* to Biel (1880), where it became one of the biggest Swiss watch factories, Omega; or the watchmaker Hahn Frères & Cie, a company which left the city in 1873 to set up in a rural village, Le Landeron, where it mechanized and specialized in the production of ébauches. Finally, the Schwob families who, even if established at La Chaux-de-Fonds as watch traders, set up their factory in the countryside, with the creation of Tavannes Watch Co. (1895).

Despite this unfavorable atmosphere, the emergence, then the spread of modern watch factories can be observed at La Chaux-de-Fonds at the end of the 19th century, under the decisive action of Jewish watchmakers.[85] Mostly from the neighboring French Alsace, they settled in La Chaux-de-Fonds during the second part of the 19th century and soon became very active in the watch trade business. Even if the Jewish population at La Chaux-de-Fonds grew during the second part of the 19th century, it was always a small minority. Indeed, its percentage of the local population went from 1.6% in 1860 to a maximum of 2.5% in 1900, when the Jewish community amounted to 914 persons.[86] Facing discrimination, this minority group had no access to the établissage networks. As a result, the Jewish milieu of La Chaux-de-Fonds got involved in the establishment of new, independent companies, sometimes gathering together several families of the community. However, their main characteristic was that they were in favor of the modernization of production modes and they were largely responsible for the first mechanized watch firms at la Chaux-de-Fonds. A census organized in December 1922 by the Chambre Suisse d'Horlogerie shows that only one of the top ten watch factories in La Chaux-de-Fonds was not directed by a Jew (Table 5): it was the Record Watch Company Founded in 1903 in the village of Tramelan. That firm was equipped with machine tools for the serial production of ébauches. It merged into a new company in 1915, grouping together the Record ébauche plant and the

Table 5: Watchmaking enterprises at La Chaux-de-Fonds employing 50 persons or more, December 1922

Enterprise	Employees	Owner
Schwob Frères & Cie	1005	Family Schwob
Fabrique Movado	195	Family Dietesheim
Record Watch Co	174	Family Perrenoud
Braunschweig Election	108	Family Braunschweig
Hirsch Fils de A. (Invar)	102	Family Hirsch
Marvin Watch SA	95	Family Didisheim
Ditisheim Paul SA	94	Family Ditisheim
Schmidt Vve Léon	70	Family Bloch, Swiss Bank Corporation
Schild & Co	60	Family Schild
Picard & Hermann Fils	50	Families Picard and Hermann

Source: MIH, Archives de la Chambre suisse d'horlogerie (CSH), Monthly statistics.

watchmaking workshop of Zélim Perrenoud at La Chaux-de-Fonds. It also had a branch at London for the sales in the British Empire.[87]

Moreover, Jewish watchmakers were part of international family and global business networks which helped them to benefit from efficient sales channels. They also had a dramatic influence outside La Chaux-de-Fonds. They created many companies in the surrounding rural areas where the watch industry was not present. Thus, they played a key role in the geographical diffusion of watchmaking within the Jura Mountains. This extension strategy into rural areas made it both possible, on the one hand, to set up modern factories in an environment not adverse to industry; and, on the other hand, to employ cheap workers who were not unionized.[88] The case of the Tavannes Watch Co. factory, mentioned above, is a good example. Another example is the company Léon Lévy & Frères, established in the city of Biel at the beginning of the 1880s, which in 1896 bought up the firm L. Gorgé & W. Rougemont, at Moutier. In the neighboring town of Delémont, one should mention the activities of Marcellus Nordmann, who registered his watchmaking firm at La Chaux-de-Fonds in 1883. He invested in an ébauche factory, the Société d'horlogerie SA, founded at Biel in 1880 and transferred to Delémont in 1883. It was then disbanded in 1884 and replaced by a new company, the Fabrique d'horlogerie de Delémont SA, in which Nordmann was a member of the Board of Directors until 1885.[89] The Picard family also invested at Delémont. Armand Picard, watchmaker at La Chaux-de-Fonds, went into partnership in 1890 with Emile Maître and founded an ébauche factory, disbanded the following year. Some years later, Armand Picard was back with his brothers Edmond and Gabriel, with whom he was associated in a watch company at La Chaux-de-Fonds. Together, in 1895 they invested in a new ébauche factory at Delémont, the company Weber, Ruedin & Cie.[90] Finally, the Blum family was also present in the small town of Delémont in the 1890s. Martin Blum established himself there in 1892, after having been associated with his brother Jules Blum, one of the wealthiest gold watch case makers of La Chaux-de-Fonds. They were, by the way, the brothers of Eugène Blum, who founded the watch company Ebel in 1911. Martin Blum opened a watch case workshop at Delémont, probably to produce cases for the family business with cheap labor. He stayed there for around ten years, then came back to La Chaux-de-Fonds where he pursued his activities as a watch case maker and then a watch trader (1906).[91]

Birth of the machine tools industry

The machine tool industry grew up during the 1880s–1900s. It was directly linked to the new needs of watchmakers and in the beginning mainly relied on mechanics from outside the Jura Mountains. At first the machine tool makers set up outside the traditional watchmaking areas; the main enterprises founded during this period were concentrated at Moutier, in the valley of Tavannes, and in the city of Biel, with the successive foundation of the firms Hauser (1898), Bechler (1904), Mikron (1908), Petermann (1914), Burri (1914), Schaüblin-Villeneuve (1915), Kummer (1917) and Tornos (1918).

The beginnings of this industry were full of difficulties. The careers of the first three main innovators (Schweizer, Laubscher and Junker) is characterized by multiple attempts to create new enterprises, technical innovation and the lack of capital, which explains the slow and nonlinear development of this industry.[92] Nicolas Junker (1851–1907) embodies this pattern quite well.[93] He was a mechanic at Schaffhausen, in the North-East of Switzerland, close to Zurich, and settled at Moutier in 1881 where he began work in an ébauche factory, the Société industrielle, founded in 1873 and subcontracting for the French firm Japy. It was then a modern factory with available capital – it was linked to the local bank Klaye, Chodat & Cie – and where Junker probably worked supervising the mechanization. As he wanted to launch his own production of machine tools, he set up an independent firm in 1883 but soon faced a lack of capital which made it impossible to grow. Then he created a new company with a local wealthy entrepreneur, Anselme Marchal, who owned a glass manufacture, and then relied on bank credit from 1886 on. In 1902, he joined with his son and the chief mechanic of his firm to form a new company. Eventually, the firm could not overcome all these financial difficulties: it was bought up by the son of Junker in 1904 and went bankrupt the following year. After this failure, Junker and his son left Moutier and established themselves respectively at Geneva and in Belgium.

Nevertheless, despite this economic failure, the enterprise of Junker appears to have been a technical success, as this factory, which employed up to around 50 persons, was a kind of hub and gave the opportunity to other mechanics to acquire knowledge which allowed them to set up their own enterprises. It was especially so in the case of André Bechler. He entered Junker's company as a mechanic after having graduated from the Technicum school of Biel, and set up his own firm in 1904, when Junker's one was failing. Bechler's company had a quick growth thanks to the development of automatic lathes designed by Junker, and it too became a business incubator: Henri Mancia, future director of Tornos; Hermann Kummer, founder of a machine company at Tramelan; Georges Cuttat, director of Manurhin, Geneva; as well as Pierre Bergonzo, director of Tanex, Geneva; all were mechanics at Bechler's company. As for the Junker's plant, it was taken over in 1911 by three associates who created the Fabrique de machines Moutier, Boy de La Tour & Cie. It changed name to Tornos in 1918, under the direction of Willy Mégel and Henri Mancia. Also, two mechanics from the neighboring village of Court, Lardon and Marchand, bought up several machines and tools from Junker's plant, and in 1906 they set up the company Ultra.

Finally, the watchmakers themselves did not play a passive role in the development of machine tools for their own needs. They were not only customers but also participated in its growth. It was particularly the case of the main watchmaking manufacturers, which used machines intensively from the 1880s, that led to the emergence of machine workshops within the watch factories, as happened for example at Longines in the 1890s. In some cases, as at the Tavannes Watch Co, the development of these workshops gave birth to spin-off companies, with the creation of Tavannes Machines SA in 1938. The owners of small firms were also engaged in this process, with collective investments into machine firms, such as the Usine mécanique de La Chaux-de-Fonds SA (1880) or the Manufacture jurassienne de machines SA, at Biel (1917).[94]

However, the involvement of watchmakers in the mechanization of their business primarily occurred through watchmaking schools, whose curricula underwent major changes at the end of the 19th century.

The modernization of watchmaking schools

Mainly founded in the 1860s and 1870s with the aim to reassert the value of traditional know-how, watchmaking schools had their mission profoundly altered at the end of the century.[95] They became key places where new technical knowledge was transmitted to new generations, trained to take over the modernization of watchmaking firms.

Soon after he came back from Philadelphia, the engineer Jacques David explained the pivotal role these institutions must have in his project to modernize the industry. At the end of his famous report, he wrote: *"The watchmaking schools and the technical drawing schools intended for apprentices must be developed; access to them must be made more and more easy for all; and then their influence will help the progress our industry has to achieve. It is necessary that students learn about the tools and the machines in use in the advanced companies I have spoken of [that is, American firms] and to give them some basic knowledge about these new manufacturing processes."*[96] In addition, David was a member of the administrative board of the Watchmaking School of Saint-Imier soon after its foundation (1868) and established himself as one of its leaders until he passed away in 1912. At his instigation, a meeting of the directors of the different watchmaking schools of the country was organized at Neuchâtel in 1877, to talk about the need *"to modify the theoretical and practical teaching methods presently used in our schools [...] as a consequence of the competition from the big American companies working with machines."*[97] In other words, the issue was to reorganize and to unify the watchmakers' training system in Switzerland, by developing teaching in mechanics and introducing machines into the schools' workshops. Eventually, nothing was decided at this meeting which mostly revealed contradictory opinions on the necessity to apply the American system to the Swiss watch industry.

The schools of Geneva, La Chaux-de-Fonds and Le Locle were the most reluctant. For them, only the promotion of high-quality, handmade watches was an effective way to compete with rival nations. For that, it was necessary to have schools supporting technical excellence and the transmission of traditional know-how. The director of the Watchmaking School of La Chaux-de-Fonds, Schouffelberger, asserted at the meeting that the administrative board of his school *"has been unanimous in deciding that no major change should be introduced into teaching methods."* According to them, the best way to compete with America was *"good watchmaking"*.[98] The administrative board of the Watchmaking School of Biel, absent at this meeting, was also very reluctant to adopt David's proposal. It wrote in the report for the period 1876–1877 that *"watchmaking being an art, it should be guided by science."*[99] However, despite this restraint, all the watchmaking schools of the country had to modernize their teaching methods, soon or later, to meet the new needs of watchmaking firms.

The main characteristic of this mutation can be observed with the diversification of professions trained within these institutions, whereas in the 1860s and 1870s they simply aimed to train watchmakers. The new professions which appeared then were mechanics and springers (*régle-uses*, women who pin the balance spring to the collet and stud, fit it to the balance staff and form the overcoil). Moreover, short courses, focusing for example on escapement planting, began to be offered and were very popular. The concentration of workers within factories continued the division and specialization of work, so many young people came to watchmaking schools to undertake short courses to become specialized workers.

The Watchmaking School of Saint-Imier was a pioneer in this mutation and exemplifies a modernization process which occurred sooner or later in all the schools of the country. In this specific case, the issue was to answer the new needs of local manufacturers, especially for the company Longines, which used the school as its main recruitment source for qualified workers. The three-year watchmaker course did not suit all the different needs. In 1881, local watchmakers asked the school to offer short training courses, which were organized by the administrative board the following year due to the *"change which has occurred during the last few years in the watchmaking production system with the creation of watch factories employing many workers on specialized parts of the fabrication."*[100] Moreover, the introduction of machines in watchmaking firms in the 1880s and the 1890s, particularly at Longines, brought about new labor needs, and it became essential to employ persons able to understand, organize and control the introduction and the use of machines within the production process. So the mechanization of watchmaking led to the training of new apprentices: mechanics. The first course in mechanics

Table 6: Training courses for the students of the watchmaking school of Saint-Imier, 1866–1916

	Creation	Number	%
Watchmaker (three years)	1866	492	50.5
Practical courses (less than three years)	1882	284	29.1
Mechanic (three years)	1896	167	17.1
Springer (one or one and a half years)	1912	32	3.3
Total		975	100

Source: *Cinquantenaire de l'École d'horlogerie de Saint-Imier*, Saint-Imier, 1916, p. 25.

was opened in 1896, *"destined to become a nursery for good workers and foremen, the need for whom can presently be seen."*[101] The number of apprentices in this area rose continuously until the beginning of the 1920s: there were 17 in 1900, 25 in 1910, and then 35 in 1920. Finally, a fourth course was offered in 1912 with the opening of a springing class. The administrative board *"decided, after the demand made by local watchmakers, to open in the spring a course for springers and to admit young girls. [...] The introduction of this course in our school is a consequence of the decrease of home work, the development of mechanized work and the lack of workers in this important part of watchmaking."*[102] The springing class gave two training courses (12 and 18 months) and consisted of about ten young girls.

All the Swiss watchmaking schools were modernized in a way similar to that seen at Saint-Imier. Some, however, took time to accept their mutation. At Biel, the school remained attached to the technical excellence model until the beginning of the 1920s. It was run by the owners of small watch firms who had no interest in new production methods, which explains why Omega had hardly any interest in the school and organized its own practical training center within the company. The Watchmaking School of Biel was finally reorganized in 1921 after the intervention of the Biel watchmakers association, which declared that *"it is the role of the school to answer the needs of factories, to populate them with qualified workers."* It deplored the fact that apprentices trained at Omega *"usually had more practical knowledge than these trained at the Watchmaking School."*[103]

The situation was similar at Geneva.[104] The school of this city had been oriented to the training of excellent watchmakers since its creation, and a workshop specializing in the repair of complicated watches was opened in 1880. Machines were introduced in the second part of the 1880s but it did not lead to a modification of the course for watchmakers. In 1891, the mechanics class was separated and became a new school of mechanics which moved away from its preoccupation with watchmaking. During the second part of the 1890s, the mechanized production of ébauches was started but it was only in 1916 that the school was reorganized with the creation of three new courses (technician, watchmaker and short course).

Finally, three new schools were opened at Porrentruy (1882), Solothurn (1884) and the Valley of Joux (1901). A unified institutional model was adopted throughout the country at the beginning of the 20th century. Having given up the ideal of technical excellence, the administrative boards proposed more open establishments, for which payment of school fees was not essential, and with the goal of offering a number of courses to meet the varied needs of industry. The watchmaking schools no longer served conservative techniques but adapted to the changing watchmaking companies.

Banks and the modernization of watchmaking

The emergence and omnipresence of big banks can be seen in many industries, as for example in textiles and the railways. However, although banks did become a major actor in the industrialization of watchmaking, their role is less obvious.

Money was a key issue in the industrial development of watchmaking, because mechanization was financially draining, and this explains the difficulties many firms faced when trying to introduce new production modes. Technical genius could not overcome everything, as Ernest Francillon experienced in the 1870s, a decade during which his company Longines was close to bankruptcy. If he eventually succeeded, it was thanks to support from private bankers.[105] The same can be seen in the case of the Omega factory of the Brandt brothers. Its growth cannot be explained just by the setting up of a modern factory and the mass production of watches, but it also required the ongoing support of the private bank Rieckel, La Chaux-de-Fonds, from the 1890s to the 1910s.[106] As for Zénith, Le Locle, the development of the factory depended on support from the Banque Cantonale de Neuchâtel, which provided loans and then became a shareholder when the company transformed into a joint-stock company (1896) and then a public company (1911). However, the presence of a bank among the shareholders does not mean that it helped an excessive growth. On the contrary, during the years 1900–1914, there were many tensions and conflicts within the management of Zénith, between Georges Favre-Jacot, main shareholder and founder of the society, who wanted unlimited growth and modernization based on bank support, and the representatives of the bank, who refused many loans and demanded a management based on depreciation and self-financing. In this particular case, the bank favored stabilization rather than industrial growth.[107] Yet, Zénith was a very uncommon case as banks as shareholders usually did not intervene in watchmaking firms before World War I. They participated in their development through loan policies, which was far less risky than direct involvement in companies without large assets[108].

The lack of available capital and big banks is a characteristic of the Jura Mountains which delayed industrialization. In an attempt to overcome this constraint, numerous small, local banks were created in all the watchmaking regions. Their main function was to collect local savings and use them to make loans to watchmakers. They were usually very small banks with limited resources, as for example the Caisse d'Epargne et d'Escompte de Saint-Imier, opened in 1877 by watchmakers of the town with a capital of 67,000 francs; or the Banque Populaire du District de Moutier SA, created in 1883 with a capital of 130,000 francs.[109] Even the private banks of Biel, Le Locle and La Chaux-de-Fonds, resulting from trade and établissage, faced severe difficulties at the beginning of the 20th century. They were taken over by the big banks after World War I.

There were eventually only cantonal banks which had enough capital to play a key role in the long run. This was especially true in the cantons of Bern and Neuchâtel, where cantonal banks massively supported watchmakers through loans. The oldest is the Banque cantonale de Berne, founded in 1834, whose first branches were opened in the watchmaking areas of the canton: Saint-Imier (1858), Biel (1858), Porrentruy (1868), Moutier (1907), Delémont (1912) and Tramelan (1921). In the neighboring canton of Neuchâtel, the creation of a cantonal bank was one of the major issues of the democratic revolution which happened in 1848, in which the watchmakers of La Chaux-de-Fonds were decisively involved.[110] The opposition to a public bank by the old aristocratic elite and private bankers of Neuchâtel postponed the realization of this project until 1883. Like in Bern, branches and agencies were soon opened in the main watchmaking towns and villages (La Chaux-de-Fonds 1883, Le Locle 1884, Fleurier 1884, Cernier 1884, Couvet 1897 and Les Ponts-de-Martel 1902). The Banque cantonale de Neuchâtel adopted an active support policy towards local industrial companies from the 1890s, by investing in firms like Zénith.

Finally, the Banque Populaire Suisse, created in the canton of Bern in 1869 under the model of a credit cooperative, played a key role in the development of watchmaking firms with its loan policies; this was one of the major reasons why it became the second largest bank in Switzerland in 1930. On the other hand, it was severely touched by the crisis of the 1930s and was saved by the federal State which invested 100 million francs in 1933.[111]

The organization of trade unions

The transformation of production in the watch industry called into question the nature of relationships between employers and workers. The appearance and spread of factory work, even within small enterprises, went with a proletarianization of labor which put an end to the anarchist movement that had been quite successful in the 1870s. Deeply linked to home work, Anarchism, unlike Marxism, stood for the suppression of the State and the *"autonomy of individuals freely associated into producer groups at the local, national and international scale."*[112] It was especially strong in the Saint-Imier area under the influence of both Adhémar Schwitzguébel and James Guillaume. Founded in 1871, the Fédération jurassienne enjoyed an international renown, but it was locally weak, having only 300 to 400 militants at its peak in the years 1873–1874, among whom were numerous Russian and German refugees. The Fédération disappeared in 1882, against the background of the restructuring of production in the watch industry, and due to an internal crisis, characterized by the gap between its political radicalization and the libertarian wishes of its members in the watchmaking region.[113]

The factory put an end to the relative autonomy of home workers. Many of them were employed by industrialists as workers and integrated within the factory, where they continued to produce parts as they had done until then in their home. Patrick Linder shed a light on this evolution in the case of Longines: this manufacturer gradually engaged its former subcontractors as factory workers, who until then had been working in independent workshops. The gathering of these workers within the factory made it possible to reorganize and mechanize work. Of course, some home workshops did survive, but they were very dependent on the big enterprises which subcontracted work to them. From then on the model of the "collective factory" (*fabrique collective*) and the relative autonomy of workers belonged to the past.

With industrialization, labor became transformed into wage-earning workers. In order to defend their working conditions, especially their wages, workers organized themselves into associations and trade unions. Some of them, usually very specialized, appeared in the late 1860s and the 1870s, as for example the associations of engravers and engine turners (1868), case makers (1871) and dial makers (1872). They had a very short existence and gave way in the 1880s and 1890s to more structured trade unions. Thus, case makers (1886), assemblers and finishers (1886), escapement makers (1887), watch jewel makers (1901), dial makers (1906) and ébauches workers (1907) set up their own associations. A huge merger movement happened before World War I, giving birth to bigger unions, as for example the Union Générale des Ouvriers Horlogers (1906).[114] Eventually, all the trade unions in the watch industry, whose leaders were then deeply influenced by revolutionary socialism, concentrated in 1911 with the creation of the Fédération des Ouvriers de l'Industrie Horlogère (FOIH). Encouraged by the failure of strikes organized in 1910, which showed the limits of a dispersed movement – as case makers were then against the centralization of trade unions – this merger strengthened the workers unions. From then on, the FOIH was organized with local sections consisting of all workers, whatever their profession, and ruled by a central committee of 13 members and local representatives in the sections. Its success was immediate: the number of members rose from 9,980 in 1912 to 17,033 in 1914. Nevertheless, it soon faced insurmountable financial difficulties due to a strike organized in 1914 at Solothurn and Waldenburg which ended with the lock-out of about 1,800 workers. This led the FOIH to approach another centralized trade union, grouping workers in the metallurgy industries. They merged in 1915 in the new Fédération suisses des ouvriers sur métaux et horlogers (FOMH).[115]

Finally, one should mention the existence of Christian trade unions, which were at odds with the FOMH because of its socialist nature, and close to the conservative party and the Catholic Church. They were supported by the owners of firms, who preferred to deal with docile trade unions with fewer protesters, rather than with an active and uncontrollable section of the FOMH.

The workers' demands varied according to their activity and their region, but a constant concern was with wages. The main objective of trade unions at the time was to maintain acceptable wages, a general aim which led sometimes to more specific, but related claims. For example, trade unions, especially the FOMH, were very reluctant, not to say openly opposed, to female workers because they created downward pressures on wages. Thus in 1917, in Biel, timers and retouchers asked the local employers association *"not to engage women on retouching after the 1st January 1918. The female retouchers currently active will be tolerated, but no others will be accepted."*[116] Such a demand was scathingly turned down, with Savoye, director of Longines, saying in 1919 that *"women workers are necessary now. It is even an absolute necessity for assembly and finishing."*[117] Indeed, watchmakers employed many women in their modernized factories. They represented a growing part of employment in the whole industry, from 34.2% of workers in 1895 to 47.4% in 1929.[118] They were not only home workers but also constituted a major part of the cheap and under-qualified labor engaged in factories.

The second category of workers who became an issue for trade unions was apprentices, not because of their direct pressure on wages, but rather because open access to the profession would bring too many workers on the market and lead to competition. Trade unions defended a restricted access to apprenticeship to reduce competition and protect wages, an objective which was completely opposed to by the employers. This was especially the reason for many conflicts and strikes among case makers at La Chaux-de-Fonds in the 1890s.

Finally, trade unions had an ambiguous position on mechanization. In the areas with an old tradition of watchmaking, such as La Chaux-de-Fonds and Le Locle, they were strongly opposed to the introduction of machines. Workers considered themselves as artisans and wanted to maintain a mode of production which allowed them to be autonomous and work within home workshops. In 1887, case makers explained that mechanization in their sector would lead to the emergence of *"a new category of unskilled workers who will take the places which belong to real workers who have completed apprenticeships."*[119] In regions where watchmaking was newly developed, the opposition of workers was not based on the use of machines per se, but rather on wages. For example, at Biel, the FOMH opposed mechanization and division of labor in the 1890s because it was accompanied by a policy of reducing wages.

Social conflicts were numerous and sometimes violent. At Longines for example, the company faced two virulent strikes, one in 1906 and the other in 1910. During the second strike, it was only with the threat of a collective lock-out of all union members in all the main factories of the region that FOMH gave up its demands. The general strike of 1918 which touched all Swiss industry soon after World War I, took place in a difficult social context. Impoverished during the war, the workers took advantage to the high economic growth of the after war period to claim for better working conditions. Fearing a communist revolution, the Swiss government, supported by the bourgeoisie, had recourse to the army and gave an ultimatum to the trade union which gave up after nearly two weeks of strike. In the watch industry, the FOMH, which had more 11,000 members in the city of Biel and the surrounding villages, was the principal intermediary and organized some local strikes. Actually, the regions where the trade union was the most powerful and the most well organized, that is, the districts of Biel (5,378 members) and Courtelary (3,197), experienced the tougher strike.[120] Work was stopped at Biel and in several factories of the villages of Saint-Imier, Villeret, Sonvilier and Renan (district of Courtelary). Elsewhere, some workshops faced abandonment of work, but closings were very rare. Despite the limited character of the strike in the Jura Mountains, which lasted only from the 9 to the 14 November, it had however an important impact on employers, as it was throughout Swiss industry. In many towns, the local authorities and leading people reacted by organizing civic guards (*garde civique*) whose aim was to combat any revolutionary movement.[121] At Tavannes, the local civic guard was directed by an executive of Tavannes Watch Co.,[122] while the industrialists and bourgeoisie of Saint-Imier gathered within a so-called "local civic union". Finally, at Le Locle, a civic guard was officially founded in January 1919 with the objective of *"ensuring the maintenance of public order, of working liberty, respect of the Constitution, of laws, individuals and property."*[123] Several of the biggest employers were among its members, such as Georges Ducommun, owner of Doxa, Georges Perrenoud, assortments (escapements) maker, and the watch maker Ariste Calame.

Despite its failure, the general strike of 1918 had direct consequences for workers in the watch industry. The first was that as the conditions of working-class life turned dramatically sour, the employers intervened and, from December 1918, decided to grant workers a pay increase. The second consequence was the reduction of working hours, a claim of the strikers. The issue of the 48-hour week, established in several other European countries, was discussed at the beginning of 1919 at the national level (the Federal Department of Public Economy and the Chambre suisse d'horlogerie). In the canton of Bern, with the biggest watch factories, the employers' association decided to reduce working hours to 52.5 hours a week (1919). Yet the trade union was not satisfied with that, and in July, the workers at several factories in Biel went on a strike which ended only in September with the acceptance of the 48-hours week by employers.[124]

A limited industrial concentration

Despite the spread of factories and mechanization, the industrial concentration process was limited in watchmaking. Geographically located in the Jura Mountains and away from the main urbanized areas of Switzerland, watch production stayed fragmented into hundreds of small production units, each specializing in only a part of the manufacturing process. Even though mechanization occurred during the period 1880–1900, there was no real concentration, either horizontally, with the merger of competing small businesses, or vertically, with the amalgamation of different stages in the watchmaking process. In 1901 official plant statistics counted a total of 663 companies active in watchmaking, employing 25,000 persons, an average of less than 40 people per enterprise.[125] Unlike the American watch industry, the Swiss remained dispersed during the 20th century. This phenomenon can largely be explained by the desire to maintain family ownership of the firms. The four biggest Swiss watch factories were, at that time, all family firms in which there was no distinction between ownership and management, despite the presence of banks among the shareholders in some cases. The Tavannes Watch Co., which was the biggest watch factory in Switzerland with 1,200 workers and a daily production of 3,200 watches in 1914,[126] belonged to the Jewish trading families Schwob, from La Chaux-de-Fonds. As for Omega (Biel), Longines (Saint-Imier) and Zénith (Le Locle), each of which employed more than 500 workers, they were all family firms. This trend was similar in the numerous small and medium sized firms throughout the Jura Mountains, which were a kind of capital for the owners' families. The absence of industrial concentration was then a direct consequence of cautious firm management, the objective of which was the conservation of family assets, together with the growth of business.

At the industry level, the absence of concentration can also be explained by macro-economic advantages in terms of competitiveness on the world market. The industrial district structure embodied a very flexible production mode which made it possible for Swiss watch industry to have an unlimited supply of a variety of products (quality, design, price, etc.),

Table 7: Firms and workers active in watchmaking and jewellery, 1901–1929

	1901	1911	1923	1929
Number of firms	663	858	972	1134
Number of employees	24,858	34,983	33,348	48,378
Average workers per firm	37.5	40.8	34.3	42.7

Source: *Feuille fédérale*, 1931, p. 193.

unlike its American competitor which focused on its domestic market and cheap, standardized watches.

Official plant statistics show that this industrial fragmentation continued during the first third of the 20th century (Table 7). The number of enterprises nearly doubled (from 663 plants in 1901 to 1,134 in 1929), as well as employees (from 24,858 persons in 1901 to 48,378 in 1929), with the average number of employees per plant staying very low (37.5 in 1905 and 42.7 in 1929). The industrial development of the Swiss watch industry was then essentially linear and followed the general trend of watch exports, whose volume rose from 8 million items in 1901 to 20.8 million in 1929. So the growth of the Swiss watch industry during this period did not lead to any concentration but rather to increased fragmentation with the emergence of new companies.

Finally, together with the absence of big business, the strong division of labor, vertically as well as horizontally, was still important in spite of mechanization. The plant statistics of 1929 reveal the specialization

Table 8: Firms and workers active in watchmaking by sector, 1929

Sector	Firms	Workers	Average workers
Watch jewels	131	3,371	25.7
Cases, gold	94	1,873	19.9
Cases, silver	34	824	24.2
Cases, other	48	2,187	45.6
Dials, crystals	83	2,369	28.5
Hands, springs, balance springs	60	1,548	25.8
Crowns, pendants, etc.	14	703	50.2
Parts, other	172	5,724	33.3
Ébauches, movements	98	6,619	67.5
Finished watches	314	20,964	66.8
Clocks	6	125	20.8
Tools	8	83	10.4

Source: *Feuille fédérale*, 1931, p. 194.

of enterprises and some trends (Table 8). Concentration of production appeared at first in ébauche and watch factories, two sectors which employed respectively on the average 67.5 and 66.8 workers, while in the areas of external parts (dials, cases and hands), enterprises were very small. The owners of these small plants are the ones who, since the 1880s, played a key role in the adoption of business agreements with the aim of defending themselves by the introduction of compulsory prices, thus compensating for their lack of competitiveness.

The persistence of numerous home workers was another feature of the limited concentration in Swiss watchmaking. Of course, the proportion of persons active in home workshops decreased, going from 87.5% of all workers in the sector in 1870 to 54.7% in 1901 (Table 9). However, despite this general trend, more than half of the workers in watchmaking at the beginning of the 20th century were active within home workshops. This persistent trait was the consequence of the economic advantages of this mode of production for *établisseurs*. Home workers lived in very poor conditions with very low incomes, which significantly contrasts with the idealized descriptions, diffused by conservative observers opposed to industrialization, such as the French sociologist Robert Pinot, who pointed out the maintenance of family life and the independence of workers.[127] These very flexible workers, who were usually not union members, were used by employers as a cyclical buffer: they got orders when the general demand was expanding and stopped receiving orders during recessions. They indeed subcontracted not only for the few modernized *établisseurs* who did not have a real factory, but also for the main manufactures of the country. In 1905, the three biggest watchmakers employed a quite high proportion of home workers: 12.8% of all employees at Langendorf SA, 17.7% at Omega and 52.8% at Longines.[128] Home work gradually lost its importance in the first

Table 9: Home workers in watchmaking, 1870–1929

	1870	1888	1901	1929
Employees, total	40,000	44,147	52,752	55,740
Home workers	35,000	32,448	28,869	8,171
As a %	87.5	73.5	54.7	14.7

Source: Scheurer, Frédéric, *Les crises de l'industrie horlogère dans le canton de Neuchâtel*, La Neuveville, 1914; and Koller, Christophe, *"De la lime à la machine". L'industrialisation et l'Etat au pays de l'horlogerie. Contribution à l'histoire économique et sociale d'une région* suisse, Courrendlin: CSE, 2003, p. 183.

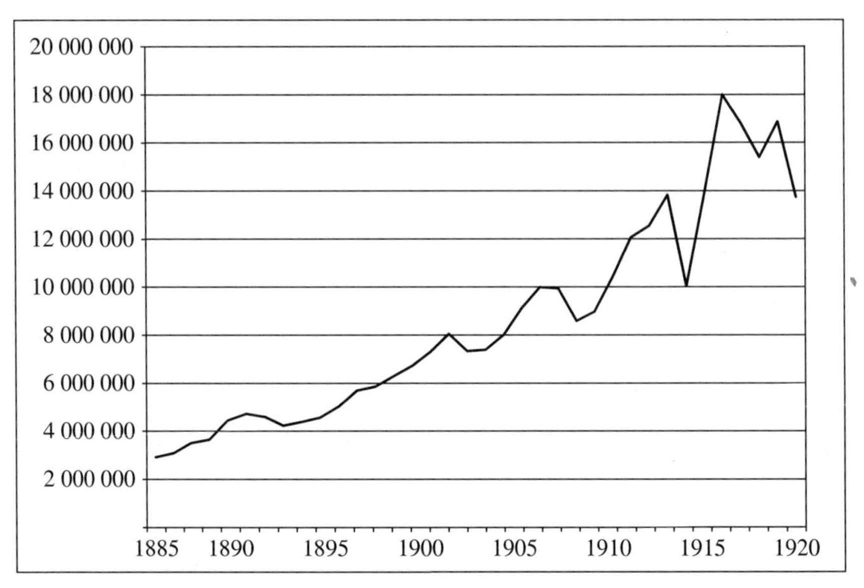

Figure 2: Export of watches and watch movements, number of pieces, 1885–1920

Source: *Statistique du commerce de la Suisse avec l'étranger*, Berne: Département fédéral des Douanes, 1885–1970.

third of the 20th century as mechanization spread. While more than half of the workers in watchmaking were home workers in 1901, there were only 14.7% in 1929. Eventually, the federal law on home working of 1940 subjected home workers to the same rules as factory workers. It then lost then its attraction and steadily disappeared.

2.3 Selling: evolution of products and markets

The years 1880–1920 were on the whole a period of high growth of watch exports and of diversification of outlets. Switzerland had a continuous increase of its watch production until World War I (Figure 2), the number of exported pieces going from 2.9 million in 1885 to 10.0 million in 1914 and peaked at 18 million in 1916. Moreover, the Swiss watch industry, whose worldwide supremacy was challenged by American competitors in the 1870s, regained its first rank at the beginning of the 20th century.[129] It then became the uncontested leader of the industry, with about 90%

of the market. America, oriented to the domestic market and to cheap, mass-produced watches, was no longer a feared rival. The growth of Swiss watch exports relied on two factors: the extension of supply and the diversification of outlets.

Indeed, mechanization made it possible for watchmakers to produce both cheap, standardized watches and luxury goods. Thus the modernization of industrial structures in watchmaking enabled diversification of supply to the world market (Table 10). The growth of exports was particularly supported by watches mass-produced in modern factories: the number of metal (steel and nickel) watches in the volume of exports rose from 20.5% in 1885 to 61.8% in 1910. Moreover, this increase was coupled with a growth in value (9.4% to 26.9%). Despite inflation, the average value of watches exported decreased; it fell from 28.1 francs in 1885 to 16.8 francs in 1900 and 12.1 francs in 1914. The development of roskopf watches was at the heart of this process. This low-value watch, characterized by a simplified movement (pin-lever escapement, 3-wheel train and large barrel) and cheap material (nickel silver cases), was marketed in 1867 by Georges-Frédéric Roskopf (1813–1889) under the name of *La Prolétaire* (the Proletarian).[130] Its production was industrialized in the 1880s–1890s, giving birth to big modernized factories, such as the company Vve Charles-Léon Schmid, at La Chaux-de-Fonds.

This growth of cheap watches occurred to the detriment of silver watches, the proportion of which fell dramatically, in terms of both volume and value. The top-of-the-range products maintained their importance. The proportion of gold watches was admittedly decreasing (16.2% in 1885 and 9.9% in 1910), but they maintained their value and in 1910, gold watches represented nearly half export value. The collapse of middle-of-the-range products (silver watches) went together with a bipolarization of the industry between bottom-of-the-range, machine produced metal watches and top-of-the-range jewelry watches.

Table 10: Structure of Swiss watches exports by kind of product, 1885–1910

	Volume %			Value %		
	1885	1900	1910	1885	1900	1910
Gold watches	16.2	12.6	9.9	41.0	40.6	45.8
Silver watches	63.3	48.6	28.3	49.6	38.2	27.3
Metal (steel and nickel) watches	20.5	38.8	61.8	9.4	21.2	26.9

Source: *Statistique du commerce de la Suisse avec l'étranger*, Berne: Département fédéral des Douanes, 1885–1910.

The evolution of products was also marked by the appearance of the wristwatch in the 1890s.[131] Initially it was only a luxury fashion product directed to female customers, and even though World War I gave certain renown to this new kind of watch, with many ordered by the belligerent armies for their soldiers, the democratization of its use did not occur until the interwar period. Thus, in 1920, wristwatches amounted to about a quarter of the production volume in Switzerland (24.6%) but they were essentially luxury products, in gold or platinum cases (51.8%). Fifteen years later, they amounted at more than two-thirds of the production (68.8%) and almost all were silver and nickel watches (95.3%).[132]

The second characteristic of this period is the diversification of outlets, which became a necessity in the 1860s and 1870s as a consequence of difficulties in the American market. This market showed a revival at the beginning of the 1880s, with watch exports amounting to more than ten million francs each year in 1880–1883; that is, about 15% of the total. Yet the end of the decade was marked by a fall and stagnation, which fluctuated around six million francs for the period 1884–1891. In 1890, the importance of the United States in Swiss watch exports amounted to only 5%.[133]

Generally speaking, the main outlets for the years 1901–1910 were Germany (22.8%), United Kingdom (14.1%), Russia (10.0%) and Austria-Hungary (10.0%); these four markets amounting to more than half of Swiss exports. Below them were twelve countries with a proportion between 1 and 10%, among whom the United States ranked seventh (5.4%).[134] A distinction can be made according to the kind of products. The main outlet for metal watches was the United Kingdom: it imported 30% of exports in 1901 and 23.6% in 1910, an importance which can be explained by the re-export of watches to the Empire and the rest of the world.[135] As for gold watches, they were mainly exported to Germany (40.7% in 1900 and 34.6% in 1910), the United Kingdom (20.0% and 15.6%) and Austria-Hungary (11.8% and 14.4%).

The beginning of mass communication

The huge expansion of Swiss watch exports at the end of the 19th century was supported by a new form of communication, the international exhibition. Together with the democratization of the use of watches, the market moved from a niche market for elite customers, characterized by individualized relationships between the maker, his agent and the final customer, to a mass consumption market. This change gave birth to new communication strategies and policies throughout the industry.

The Centennial International Exhibition at Philadelphia appears as a key moment in this evolution. It was a disaster for the image of the Swiss watch industry. The delegate Theodore Gribi was disappointed by the small number of Swiss participants and wrote that the Swiss watch industry *"drew attention by its modesty and simplicity."*[136] On the opposite side, the American industry, particularly in watchmaking, used this exhibition to show off to the world its modernism, its technological advance and the superiority of its firms. The Swiss engineers who went to Philadelphia returned home conscious of the necessity to modernize the production system. But they also knew how important it was to organize their delegations to future worldwide events. This mission was taken in hand by the Société Intercantonale des Industries du Jura (SIIJ) and its president, Ernest Francillon, director of Longines. From 1877, Francillon took charge of the organization of the Swiss delegation to the Universal Exhibition at Paris which was held in 1878. He began by supervising a preparatory exhibition in Switzerland in order to select the watchmakers who would be sent to Paris, with the aim of choosing only high quality products. In addition, Swiss watchmakers together with local authorities published an historical pamphlet on Swiss watchmaking, of which 50,000 copies were distributed in Paris.[137] Based on this experience, the SIIJ organized the delegations of watchmakers to different fairs from then on. It was particularly the case at Chicago (1893), where Swiss watchmakers, returning to the United States nearly twenty years after Philadelphia, wanted to show to the world that they had successfully overcome the American challenge.

In January 1892, the committee of the SIIJ declared that *"the Swiss watch industry must participate at the Chicago World Fair of 1893. It cannot refuse to engage in this fight, to which it is invited by the country which it counts as its strongest rival, and it must make great efforts to be represented with dignity and utility."* The selection of watches which will be shown must be organized, prohibiting 'dangerous products', defined as *"products which directly compete with American goods, that is, the movements we export to be assembled in American cases [...]. All these products (bare movements), which we send to America in a quality inferior or equal to American movements, must be mercilessly rejected by the admission jury."* So the SIIJ decided to send only the cream of Swiss watchmaking to Chicago: marine chronometers, complicated watches, precision watches with chronometer certificates, jewelry watches *"exclusively of superior quality"* and civil watches *"but exclusively those with superior quality movements and the best artistically designed cases"*.[138]

Judging by the report written in 1894 by the watchmaker Charles-Emile Tissot, official delegate of the Swiss government to Chicago, the 34 Swiss watchmakers present at the Fair had a huge success. He mentioned in particular the marine chronometers of Nardin which *"attracted the attention of all visitors and were admired by Swiss and foreign members of the jury"*[139], and the display cabinets of Patek Philippe & Cie, which *"presented to the eyes of the visitors a wonderful collection of excellently manufactured movements and watches"*. Not forgetting Ernest Francillon & Cie, of which Tissot boasted the *"quality of simple and chronograph watches and movements, with special and patented calibers [...], also a pretty collection of watches for women, richly decorated with diamonds, pearls, enamel, etc. [...]. The products of this company are universally known and rightly appreciated; they largely contribute to maintaining the good reputation of the Swiss watch."* Faced by this display of Swiss watchmaking, American makers could no longer compete. The big industrialized firm Waltham Watch *"has not made any important modifications to its calibers for some years"* and *"one finds in these products the inevitable monotony of mechanical work."* As for the pavilion of the dollar-watch factory Waterbury, *"cheap watches are piled up in a considerable number."* In conclusion, Tissot wrote that *"Switzerland has incontestably demonstrated to all that today as yesterday it occupies first place in the manufacture of watches."* He added that *"our foreign competitors should be convinced that we are strong for the fight, that we progress year by year, and that we continue to perfect the various branches of our national industry."*

The new image of the Swiss watch industry displayed at Chicago was not directed only to the American public – the American market was then declining – but to the whole world. This exhibition attracted some 26 million visitors from all countries. The challenge of Chicago was a global one: it consisted of showing to the world that Swiss watchmakers manufactured better quality watches than their American competitors; in other words that the real nature of watchmaking was Swiss. This new kind of communication through exhibitions organized at the Swiss national level went on during all the 20th century, carried on by the SIIJ which became the Chambre suisse d'horlogerie in 1899.

There were as well other new forms of communication, especially advertising, emerging within individual enterprises. Longines, Omega and Zenith were, for example, among the most innovative companies in this field. They particularly adopted publicity through posters and inserts in journals and magazines, a kind of advertisement which spread in watchmaking in the 1900s. Until World War II, the PR strategy of Swiss watch companies aimed at presenting products to the public, emphasizing notably their precision and their reliability. The principal message of these advertisements was that Swiss watches were the most accurate and thus worth being bought. It was a general communication, widespread within the industry, which reinforced the distinction between Switzerland and its rivals. If some companies introduced other elements to advertise their watches, principally fashion – for ladies watches – and sport, these themes were strongly linked to precision and basically transmitted the same message. Moreover, even if some brands, especially Rolex, became pioneers in the 1950s in introducing the idea of social distinction – the watch of presidents – they were very few, so that the motto of precision stayed the cornerstone of the advertising strategy of the Swiss watch industry until the 1970s – when the Japanese established themselves with more precise products.

2.4 Towards organized capitalism

The shift from liberal capitalism, based on individual free firms and the hands-off policy of the State, towards organized capitalism during the last third of the 19th century is a phenomenon common to all the economic systems of continental Europe and was especially pronounced in sectors which had the form of industrial districts, like the Swiss watch industry. The absence of big businesses led company owners to join together in associations and to collectively intervene in order to promote various structures conducive for the whole industry and to defend common interests.

Employers in watchmaking involved themselves in the setting up of new infrastructure necessary for the development of their businesses. Many of them were members of the boards of directors of regional railways companies and of the administrative boards of watchmaking schools. Their political engagement, usually in the progressive Radical Party, was noteworthy. This party, whose roots went back to the 1830s, gathered together particularly traders, industrialists and intellectuals opposed to the conservatives and federalists who then governed

Switzerland. Radicals engaged in the creation of a federal State in 1848 and in the social and economical transformation of Switzerland (industrialization, railways, democracy, etc.).[140] At the national, regional and local levels, watchmakers looked after their collective interest and supported a limited involvement of the State (protection of trademarks and patents, control of precious metals and professional training). At the Federal Assembly (the Swiss federal parliament), watchmakers were represented by some of the directors and owners of the biggest firms, such as Constant Dinichert (Société d'horlogerie de Montilier), Ernest Francillon (Longines), Charles-Emile Tissot (Tissot), Henri Sandoz (Tavannes Watch), Paul-Ernest Mosimann (independent watchmaker of La Chaux-de-Fonds), who were all members of the Radical Party. However, it was principally through employers' associations that their collective action took place.

The blooming of employers' associations

Several dozen employers' unions and watchmakers' associations were founded during the years 1870–1914. Their creation, often based on regional or specialty differences, corresponded to a deep mutation of the Swiss watch industry which produced new challenges that watchmakers had to face. There was at first the central question of industrialization. The introduction of machines and the quest for interchangeability of parts made it necessary to adopt common norms, especially regarding measuring units, and thus the need collaborate on a technical level. There followed the defense of the quality of watches and the promotion of their commercialization, through the introduction of precious metal control and the legal protection of trade marks. The development of watchmaking schools, patent law and the strengthening of the Swiss diplomatic network were other key issues requiring collective action. Finally, watchmakers tried to put an end to the decline in their profits by opposing wage increases claimed by trade unions and by adopting cartel agreements guaranteeing minimal sale prices. Thus, between 1880 and 1914, at least 25 cartel-like arrangements, most of them short-lived, were adopted in the watch industry.[141]

Obviously the setting up of cartels was not possible within the framework of classical liberalism. The firms' owners had to organize themselves, to negotiate and group together within associations for the defense of their common interests. Thus, various employers' associations were created in the 1880s, 1890s and 1900s, at first in some subcontracting branches, as for example the Association of Gold Case Makers of La Chaux-de-Fonds (1888), the Association of Watch Jewel Makers (1898), the Association of Hand Makers (1907), the Society of Enamel Dial Makers (1907), etc.[142] Many attempts were made to group all these associations within a collective, centralized organization but the adoption of an enduring agreement was prevented by the divergent interests of *établisseurs*, industrialists and small workshop owners.[143]

Most of the watchmakers subjected to the federal law on factories gathered together in 1887 within the Union suisse des fabriques de montres (Swiss Union of Watch Factories). However, two years later, the biggest industrialists had already left it and created their own association, the Syndicat des fabriques de montres (Watch Factories Union, 1889), presided over by the omnipresent Ernest Francillon (Longines). In 1903, this union had 28 members, employing about 7,500 workers; that is, about 30% of all people active in watchmaking. Its committee brought together the elite of the business, with Louis-Philippe Courvoisier from La Chaux-de-Fonds as a president, Jacques David (Longines), Constant Dinichert (Fabrique de Montilier), Henri Sandoz (Tavannes Watch) and Théodore Schild (Schild & Co).[144] However this association faced difficulties due to the dissidence and non-participation of some important manufactures, whose absence made it difficult to adopt common principles.

At the same time, other centralized organizations were established. A committee was created in 1888 with the aim of grouping together all the *établisseurs* under the name of the Syndicat des fabricants d'horlogerie (Watchmakers Union), but it was only successful in the cantons of Bern and Solothurn and had only 68 members (1891). As for the watchmakers of Biel, who were confronted by the claims of a well organized workers' union in the 1890s, they gathered in a local association in 1898. In 1916, the social tensions in the local factories shed light on the main problem of these numerous employers' associations: *"They do not have anybody who looks after the defense of the employers' interests, while the workers have many permanent secretaries."*[145] Thus, the employers in Biel proposed to their colleagues of the canton of Bern to leave the Watch Factories Union, which was soon dissolved, and to join together within a new, better structured and organized employers' association, the Association cantonale bernoise des fabricants d'horlogerie (Association of Watchmakers of the Canton of Bern). It led to the reconciliation of the directors of Longines and Tavannes Watch Co., located in the countryside, with their fellows in Biel, particularly the owners of Omega. These three enterprises federated all the watchmakers of the canton of Bern within an association which they controlled until the middle of the 1930s. It played a key role in the setting up of the cartel in the 1920s: both of its first permanent secretaries, the lawyers Frédéric-Louis Colomb and Lucien Clerc, left the association to take positions as directors of the new Fédération suisse des fabricants d'horlogerie (Federation of the Swiss Watch Industry, FH), to the development of which they actively contributed. Yet, until the 1920s, the main central employers' association was the Société Intercantonale des Industries du Jura (SIIJ), founded in 1876.

Table 11: Persons attending the inaugural meeting of the Société Intercantonale des Industries du Jura, 30 April 1876

Name	Canton	Function
Constant Bodenheimer	Bern	Member of the Council of State, Radical Party
Ernest Francillon	Bern	Watchmaker (Longines), Saint-Imier
Charles Lehmann	Bern	Watchmaker, Biel
Louis Müller	Bern	Watchmaker, Biel
Philippe Brémond	Geneva	Music box maker
Guedin-Chantre	Geneva	Jewel maker
Moïse Vautier	Geneva	Member of the Council of State, Radical Party
Laurent Karcher	Geneva	Trader
Hippolyte Etienne	Neuchâtel	Trader, Les Brenets
Jules Philippin	Neuchâtel	Member of the Council of State, Radical Party
Fritz Rüsser	Neuchâtel	Watchmaker, La Chaux-de-Fonds
Humbert	Solothurn	Watchmaker, Solothurn
E. Obrecht	Solothurn	Watchmaker, Grenchen
Auguste Jaccard	Vaud	Colonel and deputy (cantonal parliament)
Jules Cuendet	Vaud	Music box maker
Charles-Henri Audemard	Vaud	Watchmaker

Source: MIH, minutes of the board of the SIIJ, 30 April 1876.

The Société Intercantonale des Industries du Jura – Chambre Suisse de l'Horlogerie[146]

The Société Intercantonale des Industries du Jura (SIIJ) was created in 1876 with the aim of uniting the various local and sectorial associations in watchmaking and to defend the common interests of the whole industry to the federal government. The foundation meeting took place in April 1876 at Yverdon, to organize the delegation to Philadelphia and to react to the refusal of the federal government to appoint a representative of the watchmaking, jewelry and small mechanics sector to the consultative group to negotiate new trade agreements with Italy and France. Jules Philippin, member of the State Council of Neuchâtel (the cantonal government) and president of this first meeting, declared that *"it was essential that our industries gather into a federation and that they put themselves into a position to be listened to when the federal authority wants information or when it omits to ask for it, and to take other necessary measures."*[147]

The SIIJ quickly undertook lobbying the Federal Council, defending at the same time free trade, the limitation of taxes and the intervention of the State in favor of business. For example, it supported the restriction of postal taxes (1876, 1878), the setting up of federal legislation on precious metals control without any fiscal consequence (1877), the adoption of a liberal trade agreement with Romania (1877) and the promotion of a law on patents (1882). This political activity was reinforced by the presence of several watchmakers at the Federal Assembly as deputies, generally members of the Radical Party. In the 1880s and 1890s, when the federal customs policy turned towards protectionism, the SIJJ became a fervent defender of the *"free trade dogma."*[148] Finally, the lobbying policy organized within the SIIJ made it possible to integrate watchmaking with the other industrial and export-oriented sectors, which were grouped together in the Swiss trade and industry union (Vorort), and for which it had acted as the intermediary for watchmaking since its creation.

The SIIJ also intervened within the watch industry. It adopted a strong policy of rationalization, the objective of which was to better coordinate the production and trade activities of the industry. On the production level, its main action aimed at adopting standardized measurements. The introduction of machines and the quest for interchangeability made it necessary to adopt some common norms, especially in terms of units, and thus to collaborate on the technical level. The SIIJ played this role. Since the first meetings of the committee, Francillon made the society adopt new measurements based on the metric system (1876). The next year, this rationalization function was entrusted to a technical subcommittee whose president was Jacques David. It adopted many measures, as for example the unification of screw sizes (1879). On the commercial level, the SIIJ played a key role in the organization of the Swiss watch industry delegations to international fairs, in order to control the image of the industry in the world. It also took charge of the collection of information on foreign competitors, with the creation in 1877 of a permanent subcommittee on the United States, United Kingdom and France. Its main aim was to obtain calibers of watches produced in these countries – principally watches made by Waltham – and to lend them to its members. In 1887, this subcommittee was transformed into a business intelligence office which gave information to its members on various Swiss and foreign customers and rival companies. In 1913 this activity gave birth to an ad hoc institution, the Information Horlogère, established at La Chaux-de-Fonds.

In 1899 the SIIJ took the name of the Chambre Suisse de l'Horlogerie et des Industries Annexes (CSH). Run by a permanent general secretariat, established in the 1900s at La Chaux-de-Fonds, it was the main universal organization active in the defense of the interests of watchmaking. It continued to represent this industry at the central Swiss business association, the Vorort, and to make sure the voice of watchmaking was heard in federal discussions and negotiations related to laws on business, labor and industry, as well as other issues such as customs agreements and participation in world fairs. While many specific associations were founded in the interwar period, the CSH kept its generalist character. Even though it was officially outside the organizational structure of the watchmaking cartel, it remained the main representative of the watch industry to the federal government until its merger with the Fédération suisse des fabricants d'horlogerie (FH, 1982).

The temptation of cartels

The grouping of watchmakers within regional or sectorial associations also reveals a will to defend their material interest, through the adoption of minimum sale prices. Industrialization had led to a decrease in prices – the average value of an exported Swiss watch was halved between 1885 and 1910 – which many watchmakers tried to counteract by the adoption of cartel-like agreements. It was particularly the case among makers of parts and other subcontractors who usually possessed only small workshops and did not have the capital necessary to modernize their equipment. Gathering together within associations gave them the opportunity to attempt to impose minimum prices for selling their parts to watchmakers, thus supporting the viability of their enterprises. However, these associations were often very short-lived due to the problem of dissidence and the difficulties encountered making the many workshops respect such agreements.

One should therefore mention the original experience of the years 1906–1909, during which a price convention was implemented in all sectors and regions of the country. Weakened by the economic crisis of the years 1902–1904, the subcontractors managed to impose such an agreement. The gold case makers of La Chaux-de-Fonds led the way in 1906. Overcoming internal divisions between industrialists and artisans, they created the Société suisse des fabricants de boîtes de montres en or (Swiss Society of Gold Case Makers). It made it possible to impose price conventions on their own subcontractors (crown and pendant makers, and gold traders), and also on their customers, the gold watch makers, who then grouped in a new association including more than 400 watchmakers.[149] These agreements were about two points. First, minimum sale prices were guaranteed. Each part and each operation of the production process was endowed with a minimum negotiated price; a procedure which aimed at ensuring the profitability of the various enterprises. Second, exclusivity of business relationships between the signatories of these conventions was imposed. They forbade the delivery parts to non-signatory watchmakers, and the watchmakers agreed not to get parts from non-member workshops. This measure had the objective of excluding workshops which did not respect the minimum prices. These conventions, introduced by gold case makers, led to the adoption of similar agreements in other sectors of subcontracting: polishers, engravers, gilders, etc. who set up such cartel-like conventions during the years 1906–1907.

These conventions were essentially limited to the production of gold watches. In the field of silver and metal (steel and nickel) watches, subcontractors were too numerous and poorly organized. Also, they had to face competition from industrialists who cut prices, especially in the production of cases. It was notably the case of Omega and Tavannes Watch, who had their own case factories, respectively La Centrale, at Biel (1896) and Blanches-Fontaines, at Undervelier (1897). Engravers and case decorators, who signed an agreement for gold watches, deplored the difficulties encountered by their fellows working in silver. In a circular letter sent around 1906–1908 to all the engravers, they denounced *"the attitude of a great part of the employers in silver case decorating."*[150] The difficulties which faced silver case engravers illustrate the fragility of such agreements, which depended on the good will of signatories to follow them. To deal with various dissidence attempts, some associations in subcontracting (makers of cases, assortments, ébauches, dials, and engravers) gathered into a central association, La Vigilante, whose objective was to safeguard the conventions and to fight against *"exterior influences"*, particularly *"against groups whose interests are opposed (workers and makers)."*[151] These conventions thus appear to be a typical reaction of middle-class workshop owners against industrialization.

However, the 1906 conventions could not withstand the improvements in business conditions: watch exports grew strongly from 1908 (7.7 million pieces in 1908 and 11.7 million in 1912),[152] a growth which so favored dissidence that the conventions were torn up in 1909. However, the brief experience with the cartel-like agreements of 1906–1909 was not without long term effect. Indeed, after the euphoria of the war years, during which the watch industry saw a huge growth of business thanks to the production of war materials, it entered a period of deep industrial reorganization characterized by the adoption of a cartel agreement which would last until the middle of the 1960s.

2.5 The Swiss watch industry during World War I

World War I had a very pronounced impact on the Swiss watch industry, financially as well as industrially. The production of war material produced significant profits and had a direct influence on the modernization of the means of production.

The production of munitions

The Swiss watch industry participated in the production of munitions and war materials for belligerents during World War I. This activity, already seen in 1914, spread in 1915 with the munitions crisis faced by the Allied forces.[153] The shift to mass production of war materials in the watchmaking companies led to an ethical issue within the committee of the Chambre Suisse d'Horlogerie (CSH), the entrepreneur Louis Müller, from Biel, asserting that *"this supply will extend the war, which is not in the interests of any neutral nation."*[154] The liberal and pacifist intervention of Müller, according to whom peace is positive for the growth of business, came up against the more pragmatic attitude of his fellows. Bourquard, a watchmaker at Solothurn, summed up the general opinion quite well: *"It is true that in some factories we work night and day for the supply of war materials. Not to mention that employers and workers both get something out of it, because we receive metal, oil and many other goods necessary for our population. Besides, some work for France, others for Germany and Austria, so our neutrality does not suffer from it."*[155] The CSH did not revisit this question and watch factories worked ceaselessly producing war materials until 1918. The total amount of Swiss watch exports rose from 12.1 million francs in 1914 to more than 20 million per year during the period 1916–1918.

War production became an essential part of gross sales for some companies. This was the case, for example, of LeCoultre in the Vallée de Joux. In the 1900s the business joined with Edmond Jaeger, official supplier to the French aeronautic industry, and this led to the production of munitions and speed indicators for Allied armies. This war production amounted to 10% of the gross sales in 1915, 30% in 1916 and 68% in 1917.[156] Another particularly famous example is the company Zénith, from Le Locle. The profits from war material production made possible reduction of debt and the streamlining of the company. Indeed, the munitions' manna produced huge financial reserves (30% of paid capital in 1918) and the creation of provident funds for employees in 1915, the so-called charity fund amounting to more than 1.5 million francs in 1920.[157] Thus, the production of munitions enabled the management of Zénith to put into practice the financial policy asked for by one of its main shareholders, the Banque cantonale de Neuchâtel, which before the war had demanded debt reduction and self-financing.

Finally, at La Chaux-de-Fonds, the French business network of Jewish watchmakers seems to have been particularly important for the production of munitions for the Allied armies. It was one of these makers, Henri Picard, associated with his brothers in the society Les Fils d'Henri Picard, active in producing tools and parts for watchmakers, who played a key role. Established in Paris since 1906, where he worked for the family business, he was engaged in various industrial companies. In 1914, he was a director of a bombshell factory at Poissy (France) and thus had close connections with the French military authorities. Thanks to his relations, Picard, according to his son, *"had given the opportunity to Jules Bloch to get orders for bombshell fuses so that the watchmaking factories of La Chaux-de-Fonds worked full time during this sad period."*[158] Indeed, Jules Bloch appears to have been the key person in the subcontracting of munitions' production for Allied powers to Swiss watchmaking companies. Linked to the war material maker Schneider, at Le Creusot (France), whose agent for Switzerland he became, he acted as an intermediary with watch companies directed by his fellow Jews.[159] He was himself involved in the management of some industrial firms during the war and earned a considerable amount of money, which resulted in a sensational lawsuit for tax fraud after 1918 when the federal tax administration claimed 22.4 million francs for taxes on war profits.[160] The Jewish watchmakers were, however, not exclusively producing for Allied armies. The family Schwob (Tavannes Watch Co.) delivered bombshell fuses to all the belligerents, German as well as Allied, throughout the war.[161]

The closure of the Russian market

Finally, World War I had an important impact on commerce, with the closure of the Russian market following the Bolshevist revolution in 1917. Russia was a major outlet at the beginning of the 20th century. During the years 1908–1912, exports to that country amounted to 15 million pieces, that is 12% of all watches and movements exported.[162] Several renowned companies in Geneva and Le Locle specialized in this market; for example Paul Buhré, Ch. Tissot & Fils SA, and Tavannes Watch Co, who controlled about half of this market in 1914. For the years 1923–1936, the share of the Russian market was only 0.5% of the total of Swiss watch exports. Many companies like Tissot and Zénith had to reorientate their production towards new countries after 1917.

Tissot, who specialized in this market from the end of the 19th century, faced huge problems at the beginning of the 1920s. The total amount of its gross sales dropped from 27,000 pieces in 1920 to 11,000 in 1922, forcing the firm to fire 20% of its work force. These difficulties also led Tissot to undertake a union with the Omega.[163] As for Zenith, it also was touched by this closure. Russia represented more than half of its gross sales in the 1890s, and Zénith opened its first foreign branch in Russia in 1908.[164] Thus, despite the diversification of outlets in the 1900s, the closure of the Russian market created financial difficulties for the firm which led to a reorganization characterized by a stronger intervention by shareholders (1924).[165]

Notes

52 See the paper given by Daumas, Jean-Claude at the XIVe World Economic History Congress, Helsinki, 2006, "Districts industriels: le concept et l'histoire", <http://www.helsinki.fi/iehc2006/papers1/Daumas28.pdf> (site accessed 22 June 2009).

53 Koller, Christophe, *"De la lime à la machine". L'industrialisation et l'Etat au pays de l'horlogerie. Contribution à l'histoire économique et sociale d'une région suisse*, Courrendlin : CSE, 2003, p. 105. See also Michael C. Harrold, *American Watchmaking: A Technical History of the American Watch Industry, 1850–1930, s*upplement to the Bulletin of the National Association of Watch and Clock Collectors, vo. 14, Spring 1984.

54 Henry Bédat, Jacqueline, *Une région, une passion : l'horlogerie. Une entreprise : Longines*, Saint-Imier : Longines, 1992.

55 Koller, Christophe, *"De la lime à la machine". L'industrialisation et l'Etat au pays de l'horlogerie. Contribution à l'histoire économique et sociale d'une région suisse*, Courrendlin : CSE, 2003, p. 114.

56 Scranton, Philippe, *Endless Novelty. Specialty Production and American Industrialization, 1865–1925*, Princeton: Princeton University Press, 1997, pp. 81–107.

57 Musée international d'horlogerie, La Chaux-de-Fonds (MIH), minutes of the meetings of the Société intercantonale des industries du Jura (SIIJ), 30 June 1876.

58 *Rapport à la Société intercantonale des industries du Jura sur la fabrication de l'horlogerie aux Etats-Unis, 1876*, reprint, Saint-Imier : Longines, 1992. (English translation: David, Jacques, *American and Swiss Watchmaking in 1876*, Tasmania: Richard Watkins, 2003).

59 David, Jacques, *American and Swiss Watchmaking in 1876*, Tasmania: Richard Watkins, 2003, p. 35.

60 David, Jacques, *American and Swiss Watchmaking in 1876*, Tasmania: Richard Watkins, 2003, p. 72.

61 MIH, minutes of the meetings of the SIIJ, 3 October 1876.

62 David, Jacques, *American and Swiss Watchmaking in 1876*, Tasmania: Richard Watkins, 2003, p. 77.

63 Koller, Christophe, *"De la lime à la machine". L'industrialisation et l'Etat au pays de l'horlogerie. Contribution à l'histoire économique et sociale d'une région suisse*, Courrendlin: CSE, 2003, p. 197.

64 Kohler, Christophe, "Fabrique", *DHS*, <www.dhs.ch> (site accessed 27 September 2010).

65 Quoted by Koller, Christophe, *"De la lime à la machine". L'industrialisation et l'Etat au pays de l'horlogerie. Contribution à l'histoire économique et sociale d'une région suisse*, Courrendlin: CSE, 2003, p. 177.

66 *Feuille fédérale*, 1931, p. 193.
67 Henry Bédat, Jacqueline, *Une région, une passion: l'horlogerie. Une entreprise: Longines*, Saint-Imier: Longines, 1992.
68 *Feuille fédérale*, 1931, p. 193.
69 Ritzmann, Heiner (ed.), *Statistique historique de la Suisse*, Zurich: Chronos, 1996, p. 397; and Koller, Christophe, *"De la lime à la machine". L'industrialisation et l'Etat au pays de l'horlogerie. Contribution à l'histoire économique et sociale d'une région suisse*, Courrendlin: CSE, 2003, pp. 146 and 526–527.
70 Knobel, Joëlle, *Une manufacture d'horlogerie biennoise: la Société Louis Brandt & Frère (Omega), 1895–1935*, University of Neuchâtel, MA thesis, 1997, 136 p.; and Richon Marco, *Omega Saga*, Bienne: Fondation Adrien Brandt, 1998, 487 p.
71 Knobel, Joëlle, *Une manufacture d'horlogerie biennoise: la Société Louis Brandt & Frère (Omega), 1895–1935*, University of Neuchâtel, MA thesis, 1997, p. 89.
72 Knobel, Joëlle, *Une manufacture d'horlogerie biennoise: la Société Louis Brandt & Frère (Omega), 1895–1935*, University of Neuchâtel, MA thesis, 1997, p. 99.
73 Pasquier, Hélène, *La "Recherche et Développement" en horlogerie. Acteurs, stratégies et choix technologiques dans l'Arc jurassien suisse (1900–1970)*, University of Neuchâtel, PhD thesis, 2007, p. 353.
74 On Longines, see Henry Bédat, Jacqueline, *Une région, une passion: l'horlogerie. Une entreprise: Longines*, Saint-Imier: Longines, 1992 and Linder, Patrick, *Organisation et technologie: un système industriel en mutation. L'horlogerie à St-Imier, 1865–1918*, Université of Neuchâtel, MA thesis, 2006.
75 *Journal suisse d'horlogerie*, 1913, quoted by Gagnebin-Diacon, Christine, *La fabrique et le village: la Tavannes Watch Co, 1890–1918*, Porrentruy: CEH, 2006, p. 29.
76 Hostettler, Patricia, *Naissance et croissance d'une manufacture horlogère: la fabrique de montres Zénith SA au Locle (1865–1925)*, University of Neuchâtel , MA thesis, 1987.
77 Berthoud, Robert, *Répertoire des brevets*, s.l., s.d.
78 Quoted by Jequier, François, *De la forge à la manufacture horlogère (XVIIIe–XXe siècles). Cinq générations d'entrepreneurs de la vallée de Joux au cœur d'une mutation industrielle*, Lausanne: Bibliothèque historique vaudoise, 1983, p. 295.
79 Evard, Maurice, "Fontainemelon", *DHS*, <www.dhs.ch> (site accessed 26 June 2009).
80 Archives de l'Etat de Neuchâtel (AEN), Department of Industry, Control of factories, 363–368.
81 Knoepfli, Adrian, "International Watch Co. (IWC)", *DHS*, <www.dhs.ch> (accessed 10 October 2010).
82 Harrold, Michael C., *American Watchmaking. A Technical History of the American Watch Industry, 1850–1930*, Columbia: NAWCC, 1984, p. 33.
83 Donzé, Pierre-Yves, *Les patrons horlogers de La Chaux-de-Fonds (1840–1920). Dynamique sociale d'une élite industrielle*, Neuchâtel: Alphil, 2007.
84 MIH, Swiss Watch Industry Yearbook, 1880 and 1900.
85 Donzé, Pierre-Yves, *Les patrons horlogers de La Chaux-de-Fonds (1840–1920). Dynamique sociale d'une élite industrielle*, Neuchâtel: Alphil, 2007.
86 Donzé, Pierre-Yves, *Les patrons horlogers de La Chaux-de-Fonds (1840–1920). Dynamique sociale d'une élite industrielle*, Neuchâtel: Alphil, 2007, p. 61.
87 *Feuille officielle suisse du commerce*, various years.

88 Kohler, François, “Les communautés juives dans le Jura (xixe–xxe siècles)”, *L’Hôtâ*, n° 20, 1996, pp. 73–84.

89 Archives du registre du commerce, Delémont.

90 Kleisl, Jean-Daniel, *Le patronat de la boîte de montre dans la vallée de Délémont: l’exemple de E. Piquerez S.A. et de G. Ruedin S.A. à Bassecourt (1926–1982),* Delémont: Alphil, 1999, pp. 39–41.

91 Donzé, Pierre-Yves, *Les patrons horlogers de La Chaux-de-Fonds (1840–1920). Dynamique sociale d’une élite industrielle*, Neuchâtel: Alphil, 2007.

92 Marti, Laurence, “Le tour à poupée mobile. (Jura suisse, 1870–1920)” in Belot Robert, Cotte Michel and Lamard Pierre (ed.), *La technologie au risque de l’histoire*, Belfort-Montbéliard: UTBM, 2000, pp. 191–198.

93 Marti, Laurence, “Nicolas Junker, Fabrique de machines, Moutier (1883–1905) ou les difficultés d’une entreprise innovante à la fin du XIXe siècle”, in Tissot, Laurent (ed.), “Entreprises et réseaux. Les acteurs de l’industrialisation dans l’Arc jurassien”, *Actes de la Société jurassienne d’Emulation*, 1999, pp. 298–305.

94 *Feuille officielle suisse du commerce*, various years.

95 Simonin, Antoine, and Fallet, Estelle (ed.), *Dix écoles d’horlogerie suisses : Chefs-d’œuvre de savoir-faire*, Neuchâtel : Editions Simonin, 2010.

96 David, Jacques, *American and Swiss Watchmaking in 1876*, Tasmania: Richard Watkins, 2003, p. 75. (English translation of David, Jacques, *Rapport à la Société intercantonale des industries du Jura sur la fabrication de l’horlogerie aux Etats-Unis, 1876*, Saint-Imier : Edition des Longines, 1992.)

97 Mémoires d’Ici, Saint-Imier (MDI), Papers of the Watchmaking School of Saint-Imier, report on a meeting of the directors of Swiss watchmaking schools, 17 May 1877.

98 MDI, Papers of the Watchmaking School of Saint-Imier, report on a meeting of the directors of Swiss watchmaking schools, 17 May 1877.

99 Annual report of the Watchmaking School of Biel, 1876–1877, p. 6.

100 Archives de l’Etat de Berne (AEB), BB IV 1121, letter of the administrative board to the Direction of Interior Affairs, 23 May 1882.

101 AEB, Annual report of the Watchmaking School of Saint-Imier, 1897–1898, p. 11.

102 AEB, Annual report of the Watchmaking School of Saint-Imier, 1911–1912, p. 7.

103 Archives of the Association of watchmakers of the canton of Bern, minute of a meeting between the Society of watchmakers of Biel and the administrative board of the Watchmaking School of Biel, 26 February 1921.

104 *Les écoles suisses d’horlogerie*, Zurich: Fritz Lindner, 1948, pp. 115–119.

105 Marti, Laurence, *Une région au rythme du temps. Histoire socio-économique du Vallon de Saint-Imier et environs, 1700–2007*, Saint-Imier: Edition des Longines, 2007, pp. 126–127.

106 Knobel, Joëlle, *Une manufacture d’horlogerie biennoise: la Société Louis Brandt & Frère (Omega), 1895–1935*, University of Neuchâtel, MA thesis, 1997.

107 Hostettler, Patricia, *Naissance et croissance d’une manufacture horlogère: la fabrique de montres Zénith SA au Locle (1865*–1925*),* University of Neuchâtel , MA thesis, 1987.

108 Perrenoud, Marc, “Crises horlogères et interventions étatiques: le cas de la Banque cantonale neuchâteloise dans l’entre-deux-guerres”, in Cassis Youssef and Tanner Jakob (ed.), *Banken und Kredit in der Schweiz, 1850–1930*, Zurich: Chronos, 1993, p. 211.

109 *Feuille officielle suisse du commerce*, 1883.

110 Donzé, Pierre-Yves, *Les patrons horlogers de La Chaux-de-Fonds (1840–1920). Dynamique sociale d'une élite industrielle*, Neuchâtel: Alphil, 2007.

111 Baumann, Jean-Henning, "Banque populaire suisse (BPS)", *DHS*, <www.dhs.ch> (site accessed 24 June 2009).

112 "Anarchisme", *DHS*, <www.dhs.ch> (site accessed 24 June 2009).

113 Kohler, François, "Fédération jurassienne", *DHS*, <www.dhs.ch> (site accessed 24 June 2009).

114 Beck, Renatus (ed.), *Voies multiples, but unique. Regard sur le syndicat FTMH 1970–2000*, Lausanne: Payot, 2004, pp. 134–36.

115 Gerber, Jean-Frédéric, "Le syndicalisme ouvrier dans l'industrie suisse de la montre de 1880 à 1915", in Gruner, Erich, *Arbeiterschaft und Wirtschaft in der Schweiz, 1880–1914*, Zurich: Chronos, 1988, vol. 2, pp. 479–528.

116 Archives of the Association of watchmakers of the canton of Bern, minutes of the board, 13 December 1917.

117 Archives of the Association of watchmakers of the canton of Bern, minutes of the board, 25 September 1919.

118 Koller, Christophe, *"De la lime à la machine". L'industrialisation et l'Etat au pays de l'horlogerie. Contribution à l'histoire économique et sociale d'une région suisse*, Courrendlin: CSE, 2003, p. 202.

119 SSA, FTMH, 04–0377 Société suisse des patrons monteurs de boîtes, minutes of the board, 22 October 1887.

120 Kohler, François, "La grève générale dans le Jura", in Vuilleumier Marc (ed.), *La grève générale de 1918 en Suisse*, Genève: Grounauer, 1977, pp. 61–78.

121 Guex Sébastien, "À propos des gardes civiques et de leur financement à l'issue de la Première Guerre mondiale", in *Pour une histoire des gens sans histoire : ouvriers, exclues et rebelles en Suisse, XIXe–XXe siècles*, Lausanne : Editions d'en bas, 1995, pp. 255–264.

122 Gagnebin-Diacon, Christine, *La fabrique et le village: la Tavannes Watch Co, 1890–1918*, Porrentruy : CEH, 2006, p. 93.

123 Archives communales du Locle, Statutes of the Civil Guard of Le Locle, art. 2, 21 January 1919.

124 MDI, Pfister report, 1919.

125 *Feuille fédérale*, 1931, p. 193.

126 Gagnebin-Diacon, Christine, *La fabrique et le village: la Tavannes Watch Co, 1890–1918*, Porrentruy: CEH, 2006.

127 Pinot, Robert, *Paysans et horlogers jurassiens*, Genève: Gronauer, 1979, 352 p.

128 Landes, David S., *Revolution in time, clocks and the making of the modern world*, Cambridge: Harvard University Press, 1983, p. 383.

129 Landes, David S., *Revolution in time, clocks and the making of the modern world*, USA: Harvard University Press, 1983, pp. 347–354.

130 Buffat, E, *History and design of the Roskopf watch*, Australia: Richard Watkins, 51 pp, available from <www.watkinsr.id.au> (site accessed 20 September 2010).

131 Béguelin, Sylvie, "Naissance et développement de la montre-bracelet: histoire d'une conquête (1880–1950)", *Chronometrophilia*, 37 (1994), pp. 33–43.

132 Ritzmann, Heiner (ed.), *Statistique historique de la Suisse*, Zurich: Chronos, 1996, p. 627.
133 Koller, Christophe, *"De la lime à la machine". L'industrialisation et l'Etat au pays de l'horlogerie. Contribution à l'histoire économique et sociale d'une région suisse*, Courrendlin: CSE, 2003, p. 114, and own calculation.
134 Fallet-Scheurer, Marius, *Le travail à domicile dans l'horlogerie suisse et ses industries annexes*, Berne: Imp. de l'Union, 1912, p. 69.
135 Fallet-Scheurer, Marius, *Le travail à domicile dans l'horlogerie suisse et ses industries annexes*, Berne: Imp. de l'Union, 1912, p. 69.
136 Quoted by Koller, Christophe, *"De la lime à la machine". L'industrialisation et l'Etat au pays de l'horlogerie. Contribution à l'histoire économique et sociale d'une région suisse*, Courrendlin: CSE, 2003, p. 288.
137 Koller, Christophe, *"De la lime à la machine". L'industrialisation et l'Etat au pays de l'horlogerie. Contribution à l'histoire économique et sociale d'une région suisse*, Courrendlin: CSE, 2003, p. 291.
138 MIH, minutes of the board of the SIIJ, 18 January 1892.
139 Tissot, Charles-Emile, *Rapport spécial sur l'exposition d'horlogerie*, s.l.: s.n., 1894, 62 p.
140 Humair, Cédric, *1848 : Naissance de la Suisse moderne*, Lausanne : Antipodes, 2009.
141 Humair, Cédric, *Développement économique et Etat central (1815–1914): un siècle de politique douanière suisse au service des élites*, Berne: Lang, 2004, p. 354.
142 *Feuille officielle suisse du commerce*, various years.
143 Gerber, Jean-Frédéric, "Le syndicalisme ouvrier dans l'industrie suisse de la montre de 1880 à 1915", in Erich Gruner, *Arbeiterschaft und Wirtschaft in der Schweiz, 1880–1914*, Zurich: Chronos, 1988, vol. 2, pp. 479–528.
144 *Feuille officielle suisse du commerce*, 1903.
145 Archives of the Compagnie des Montres Longines Francillon SA, report given by the committee of the Society of Watchamkers of Biel to the constitutive meeting of the Association of Watchmakers of the Canton of Bern, 11 April 1916.
146 MIH, minutes of the board of the SIIJ, 1876–1920.
147 MIH, minutes of the board of the SIIJ, 30 April 1876.
148 Humair, Cédric, *Développement économique et État central (1815–1914): un siècle de politique douanière suisse au service des élites*, Berne: Lang, 2004, p. 352.
149 MIH, *Statuts et conventions du 11 mars 1906*, La Chaux-de-Fonds: Imp. du National suisse, 1906.
150 Schweizerische Wirtschafts Archiv, Bâle (SWA), H 45, "Programme d'activité à déployer pour la boîte argent", s.d. [1906–1908].
151 SWA, H 45, report of the 17 November 1908.
152 *Statistique du commerce de la Suisse avec l'étranger*, Berne: Département fédéral des Douanes, 1908–1912.
153 Luciri, Pierre, "L'industrie suisse à la rescousse des armées alliées. Un épisode de la coopération inter-alliée pendant l'été 1915", *Relations internationales*, 1974, pp. 99–114.
154 MIH, minutes of the board of the CSH, 28 April 1915.
155 MIH, minutes of the board of the CSH, 28 April 1915.

156 Jequier, François, *De la forge à la manufacture horlogère (XVIIIe–XXe siècles). Cinq générations d'entrepreneurs de la vallée de Joux au cœur d'une mutation industrielle*, Lausanne: Bibliothèque historique vaudoise, 1983, p. 433.

157 Hostettler, Patricia, *Naissance et croissance d'une manufacture horlogère: la fabrique de montres Zénith SA au Locle (1865–1925),* University of Neuchâtel , MA thesis, 1987.

158 MIH, letter from Henri Picard fils to J.-P. Chollet, 2 May 1972.

159 According to Luciri, Pierre, "L'industrie suisse à la rescousse des armées alliées. Un épisode de la coopération inter-alliée pendant l'été 1915", *Relations internationales*, 1974, p. 113.

160 Perrenoud, Marc, "L'évolution industrielle de 1914 à nos jours", in *Histoire du Pays de Neuchâtel*, Hauterive: Gilles Attinger, vol. 3, 1993, p. 147.

161 Gagnebin-Diacon, Christine, *La fabrique et le village: la Tavannes Watch Co, 1890–1918*, Porrentruy: CEH, 2006, pp. 45–47.

162 Koller, Christophe, *"De la lime à la machine". L'industrialisation et l'Etat au pays de l'horlogerie. Contribution à l'histoire économique et sociale d'une région suisse*, Courrendlin: CSE, 2003, p. 431.

163 Pasquier, Hélène, *La "Recherche et Développement" en horlogerie. Acteurs, stratégies et choix technologiques dans l'Arc jurassien suisse (1900–1970)*, University of Neuchâtel, PhD thesis, 2007, p. 41.

164 Hostettler, Patricia, *Naissance et croissance d'une manufacture horlogère: la fabrique de montres Zénith SA au Locle (1865–1925),* University of Neuchâtel , MA thesis, 1987, p. 49.

165 Hostettler, Patricia, *Naissance et croissance d'une manufacture horlogère: la fabrique de montres Zénith SA au Locle (1865–1925),* University of Neuchâtel, MA thesis, 1987.

CHAPTER 3
The watchmaking cartel (1920–1960)

During the postwar euphoria watch exports grew from 13.8 million pieces in 1913 to 16.9 million in 1919. But in 1920–1922 the Swiss watch industry experienced a deep crisis. Watch exports suddenly dropped from 13.7 million in 1920 to 7.9 million in 1921 and 9.6 million in 1922, causing a large increase of unemployment. There were more than 30,000 jobless in watchmaking during the summer of 1921,[166] and the Longines factory, one of the biggest in the country, fired 40% of its employees between 1918 and 1921.[167] This crisis led to a shared will to profoundly reorganize the sector.

The consequence of these economic difficulties was a general decrease of prices, which in turn strengthened the crisis, with pressure on salaries, conflicts with trade unions, lack of liquid assets in companies and bankruptcies. Above all, the drop in prices affected subcontractors, who tried to sell off their stock and production, sometimes at a loss, with the aim of creating cash and avoiding bankruptcy. The disastrous effects of the competition between subcontracting workshops played a key role in the awakening of the necessity to adopt a corporatist policy which could help maintain them.

The 1920–1922 crisis had a second negative effect on the Swiss watch industry: it favored the emergence of competitors in other countries through the emigration of Swiss watchmakers, driven abroad by economic hardships. This phenomenon reinforced the feeling of the necessity to protect the watch industry. Seeing that *"there is for our national industry a serious danger from foreign competition growing and durably establishing itself"*[168], in 1922 the Federal Council (federal government) decided to allocate five million francs in aid, about 3% of the 1921 export value of watches. This aid was granted to watchmaking enterprises in order to help them export with competitive prices and maintain the industry in Switzerland. Obviously, the Swiss government also had the objective of relieving some banks, particularly the cantonal ones, which faced difficulties owing to the loans granted to watchmakers. The companies which received this federal support had to make the commitment to *"supply goods whose origin and makers were Swiss."*[169] The struggle against the transfer of watchmaking technologies to other nations and the desire to maintain the industrial structure based on small and medium sized firms, were the two main objectives of the reorganization of watchmaking in the form of a cartel during the interwar years. These were certainly old problems, which watchmakers had been talking about since the 1870s, but they became hot issues after World War I.

3.1 The problem of *chablonnage* and the struggle against industrial transplantation

The basic problem the Swiss watch industry had to confront during the interwar years was the development of what was called *chablonnage*; that is, exporting disassembled watches (movements or movement parts) and assembling them in the countries in which they were sold.[170] The main purpose was to avoid paying the high customs duties on finished watches. Yet, what really scared watchmakers was that the techniques and know-how transferred through *chablonnage*, within assembly workshops set up abroad, would lead to the emergence of new rival firms, so challenging the dominant position of Switzerland. The rise of customs protectionism after World War I boosted *chablonnage* and made Swiss watchmakers aware of the need to take steps to put an end to such practices.

On the basis of Swiss foreign trade statistics, it is possible to determine the part played by *chablonnage* in watch exports (Figure 3) and to evaluate its spread after World War I. Up until 1914, Swiss watchmakers were not unduly concerned about the practice. Exports of movements showed a steady increase, rising from 297,000 units in 1890 to 1.2 million in 1914. In relative terms, however, this growth was not that significant: movements as a share of Swiss watchmaking exports (number of units) rose until 1906 (13.6% as against 5.9% in 1890), then fell during the years leading up to the war. Likewise, the spread of mechanized production at the beginning of the 20th century made interchangeability of parts possible and facilitated the export of disassembled watches, as assembling no longer required fitting: *"before the war, we exported chablons [sets of parts ready to be assembled] only after having assembled the watches beforehand to check that they worked. The watches were then disassembled and the chablons exported."*[171] After the war, however, movement exports began to pose a problem. Such exports not only rose sharply in

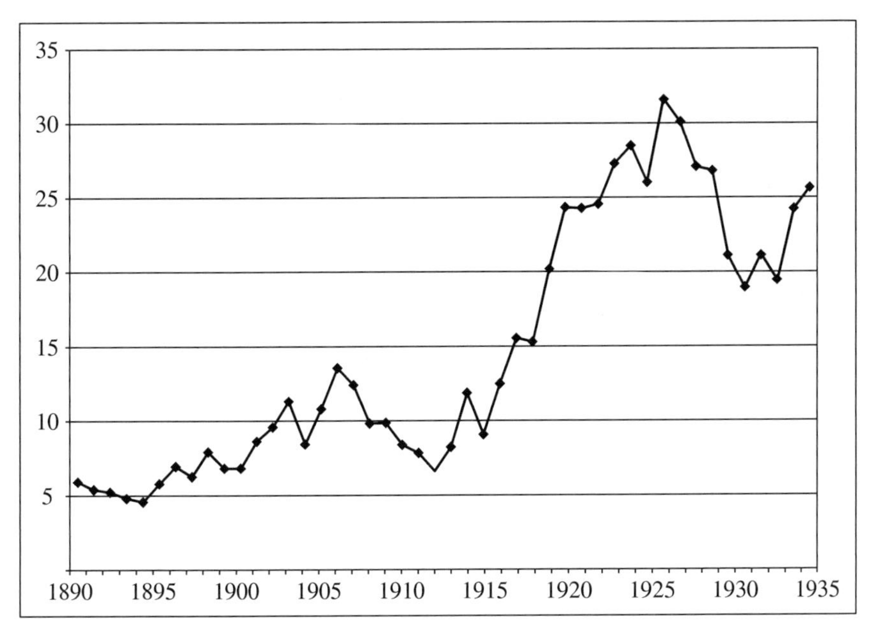

Figure 3: Swiss exports of movements as a percentage of total exports, 1890–1935 (number of units)

Source: *Statistique du commerce de la Suisse avec l'étranger*, Berne : Département fédéral des Douanes, 1890–1935.

absolute terms (2.4 million units in 1918, peaking at 5.6 million units in 1929) but above all tended to become a dominant practice in exports (their relative share of watchmaking exports went from 11.9% in 1914 to a high of 31.6% in 1926). In general, movements averaged 25.1% of watchmaking exports from 1920 to 1935, as compared with 11.5% for 1900 to 1920.

In fact, *chablonnage* was limited to a small number of countries up until the mid-1920s, when it became more widespread as a result of customs protectionism (see Table 12). Before 1930, North America (USA and Canada), Germany, Russia and Japan absorbed nearly 90% of Swiss exports of watch movements. They were the main outlets of Swiss watch industry.

Germany in particular was a long-standing outlet for Swiss watchmakers. The development of its own watch industry occurred due to strong links with its Swiss neighbor. After World War I, the export of *chablons* to Germany increased, especially to the city of Pforzheim, where the local jewelry industry was facing crisis and unemployment; it then diversified into watch case making and the assembly of watches. In 1928, the number

Table 12: Main destinations of movement exports for Swiss watches, 1900–1930

	1900	1910	1920	1930
Movements exported, no. of units	498,892	873,522	3,340,982	3,421,959
USA (%)	40.7	29.1	70.3	36.3
Russia (%)	15.3	21.4	–	–
Japan (%)	19.8	10.3	10.9	8.6
Germany (%)	9.8	7.7	–	8.7
Canada (%)	9.6	21.6	9.1	11.2
Other (%)	4.9	10.0	9.7	35.2

Source: *Statistique du commerce de la Suisse avec l'étranger*, Berne: Federal Customs Department, 1900–1930.

of workers occupied in the assembly of Swiss *chablons* at Pforzheim amounted at about one thousand persons.[172] In addition, some Swiss watchmakers contributed to the expansion of this industry in Germany. It was for example the case of the brothers Rudolf and Hermann Geering, who had owned a company at La Chaux-de-Fonds since 1905. They founded a subsidiary in Pforzheim at the beginning of the 1920s, which got parts from Switzerland and assembled them in Germany.[173] These firms, organized on an industrial scale, exported parts to Switzerland and so became competitors of Swiss subcontracting workshops. It was especially widespread in watch case making, a part the Swiss watchmakers imported in only very small quantities before World War I (16,601 items in 1910, which is 5.1% of all imported cases). In the 1920s, this practice spread and expanded, reaching 145,474 items imported from Germany in 1930, which is 27.6% of all imported cases. This phenomenon was not a problem in itself, as at the time only 1.1% of Swiss watches exported were equipped with German cases, but it made an impression and contributed to the will to control the Swiss watch industry.

The practice of *chablonnage* exports to Russia, one of the main outlets of the Swiss watch industry in the 1900s, appeared as a direct consequence of the industrial policy of local authorities who encouraged the establishment of foreign firms on their territory, thanks to a protectionist custom policy. The emigration of Swiss watchmakers to the Russian empire was important throughout the 19th century. Alain Maeder has shown that more than 400 passports had been delivered to traders and watchmakers of the canton of Neuchâtel in the years 1798–1890.[174] Among them, there were some representatives of the watch companies Paul Buhré, Moser, Zénith and above all Ch. Tissot & Fils, which made Russia their main outlet.[175] They set up branches to which part of the production process, namely assembling and casing, was gradually transferred. In 1878, an engineer from Neuchâtel established himself at St. Petersburg where he opened a watch factory in which some twenty Swiss watchmakers were working.[176] However, when putting an end to commercial relationships with Swiss watchmakers, the new Bolshevist regime ended the practice of *chablonnage* and cut itself off from an important source of know-how. Thus, when they decided to develop their own national watch industry, in the 1930s, the Soviet authorities adopted an active technology transfer policy. They bought up two American companies in 1930 (Hampden Watch Co. and Ansonia Watch Co.)[177] and several times met with Swiss watchmakers, trying to get a supply of machine-tools and parts.[178]

However, the two countries towards which *chablonnage* appears to have been problematic were the United States and Japan. The technology transfer which took place within such a practice reinforced their watch industries and contributed to making these countries the main competitors of Swiss watchmakers on the global market during the 20th century.

The United States

The main country to which disassembled watches were exported was one of the most important commercial partners of the Swiss watchmakers: the United States. While economic relations between both countries were of a relatively liberal nature during the second part of the 19th century, successively increasing tariffs were adopted at the beginning of the 20th century (1909, 1913, 1922, 1928 and 1930), under the pressure of the American watchmakers, especially the Waltham Watch Co. The election victories of the Republicans, to the Congress in 1920 and then the White House in 1928, encouraged these high customs duties. Relations between the United States and Switzerland became fraught. In April 1930, 15,000 workers marched at Biel, protesting against the new American customs tariff and asking for a boycott of American products. A new trade agreement was finally signed by Switzerland and the Unites States in 1936: it introduced a more liberal American trade policy regarding imports, but was more severe towards watch smuggling.

Smuggling was indeed very much encouraged by the American tariff barrier. It was estimated at 200,000 to 300,000 watches a year during the period 1930–1935.[179] *Chablonnage* as well was booming: the number of watch movements exported to the United States reached 203,000 pieces in 1900, 254,000 in 1910 and 2.3 million in 1920, before dropping to 1.2 million in 1930 due to the great depression. These movements were assembled and cased by American trading companies which marketed watches in North American (Bulova, Benrus, Gruen and Wittnauer).[180] Thus, *chablonnage* enabled firms specializing in distribution to diversify to production and establish themselves as important manufacturers in the United States thanks to the acquisition of know-how from Switzerland. The case of the Bulova Watch Co. is a good example of this process.

This firm was founded by a Czech immigrant, Joseph Bulova (1852–1935), who opened a watch and jewellery shop in New York in 1875 and began importing Swiss watches in 1887. A branch was opened in 1911 in Bienne, Switzerland to source Swiss products directly, which was soon turned into a watchmaking workshop. By the mid-1910s, Bulova had a dual organizational structure: it produced watches in Switzerland then marketed them in the USA. However, rising USA customs duties led the firm to transfer part of its production to America, thereby cutting the umbilical cord with Swiss production facilities. During the 1920s, it developed a strategy for acquiring specific watchmaking know-how by buying up Swiss factories manufacturing assortments and machine tools. With this newfound technical mastery, it was able to open an ébauche plant in the USA in the early 1930s. However, it continued to produce some movements on Swiss soil which were then assembled in America. In the 1930s, Bulova also distributed Swiss brands in the USA, acting as an agent for Vacheron-Constantin and taking a stake in Longines-Wittnauer, the importer and distributor of Longines watches in the US. It is not clear exactly what role these two Swiss manufacturers played in *chablonnage* exports to the USA, but it seems likely that the ties with Bulova led these two prestigious Swiss brands to assemble watches in the USA, enabling them to overcome serious financial difficulties during the depression. For Bulova, the transfer of technologies realized during the interwar period led to an industrial success which enabled the company to establish itself as an important watchmaker beside the more traditional companies like Waltham and Elgin.[181]

Technology transferred via *chablonnage* also gave the Japanese watch industry a real boost. After North America, Japan was the primary destination for Swiss watch movements. Following the Sino-Japanese war, Japan adopted a new trade policy aimed at securing a revision of the unequal treaties imposed in the mid-19th century. In the late 1890s, it opted for progressive customs protectionism designed to protect domestic infant industries, and the watchmaking sector was no exception. Watch imports were hit by a series of increases in customs duties in 1899, 1906 and 1926 aimed at guaranteeing protection for the manufacturer Hattori Kintaro on the Japanese market. After opening a watch repair shop (1877) followed by a watch shop (1881), Hattori started making clocks (1892) then pocket watches (1895). His firm, which later took the name of Seiko, dominated the Japanese watchmaking sector at the time. His products accounted for a significant share of domestic production from 1906 to 1930, with clocks representing 47.9% and watches 85.2%.[182]

Japanese customs protectionism led directly to *chablonnage*, which increased considerably in the interwar period. Movements accounted for 30.9% of the volume of Swiss watch exports to Japan from 1900 to 1915, then 42.1% from 1915 to 1925 and 80.5% from 1925 to 1940 (see Figure 4). They were imported by two types of companies and were a key driving force behind the transfer of technology.

The first type of importer consists of Japanese watch distributors. The first name that comes to mind is the firm Tenshodo, which opened in 1879 as a dealer selling luxury products from the West including Swiss watches.[183] In 1918, Tenshodo opened a branch in Switzerland, at La Chaux-de-Fonds,[184] as well as an assembly plant in Tokyo.[185] It was one of the main channels for the procurement of *chablons* in Japan up until the late 1920s. However, the company was a victim of its own policy, which led it to massively stockpile movements before the 1926 increase in customs duties, driving it into bankruptcy in 1929. After being restructured in the 1930s, it remained in business but stopped assembling Swiss watches. Tenshodo's bankruptcy directly benefited the second Japanese company active in importing *chablons*, the firm Hattori. The latter imported watch movements directly from a breakaway Swiss company and bought up Tenshodo's stockpiles when they were auctioned off in 1929.[186] Acquiring such a large quantity of these watch movements enabled Hattori to develop an aggressive commercial policy outside the Japanese empire in

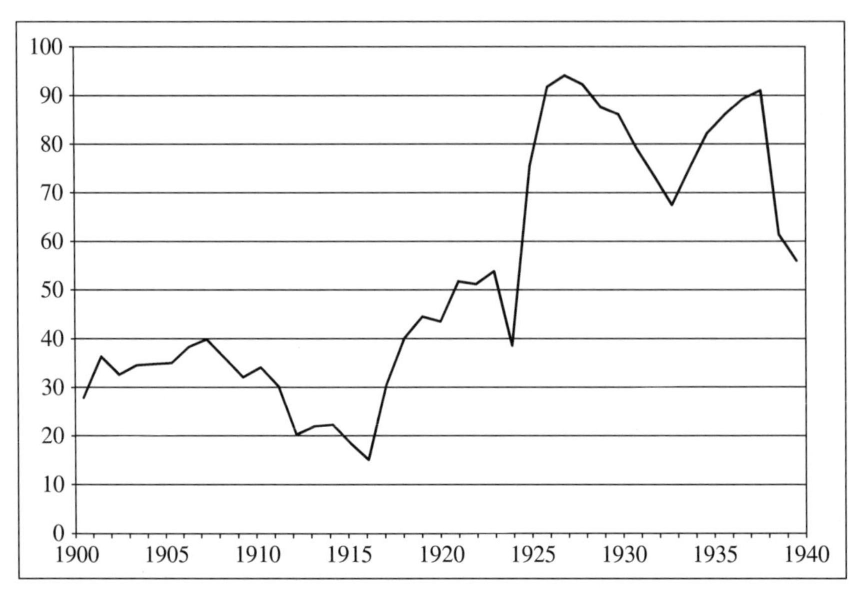

Figure 4: Swiss exports of movements as a percentage of total exports 1900–1940

Source: *Statistique du commerce de la Suisse avec l'étranger*, Berne: Federal Customs Department, 1900–1940.

the early 1930s. By assembling these movements in its own plants using watchcases it had manufactured itself, the company was able to produce "Swiss watches" at a much lower cost price than Swiss watchmakers and so challenge their ranking in certain markets, for example India.[187] In this way, *chablonnage* enabled Japanese watch merchants, in particular Hattori, to improve their ranking. Yet it was not a vector for technology transfer. Hattori had been producing watches since the 1890s, thanks to know-how acquired in the USA and Switzerland through other channels, such as visits by engineers and machine imports.[188]

The second type of company involved in *chablonnage* in Japan consisted of Swiss watch manufacturers who opened up assembly plants in the country and engaged in genuine technology transfer. In particular, this is what happened with a small-scale watch importer who settled in Yokohama in the 1890s, the Swiss Rodolphe Schmid, once his business got off the ground. He began importing movements into Japan two years after the 1906 customs hike and assembled his watches in a workshop which became a thriving undertaking. By 1920, his company had become the second largest watch manufacturer in Japan in terms of employees. With its 110 workers, it was perhaps far behind Hattori (1,943 workers) but well ahead of all other watch producers in the country.[189] In 1930 Schmid, through some of his Japanese employees, helped found Citizen Watch Co.[190] In 1933, Schmid imported machine tools for manufacturing ébauches, and in 1934 he commissioned a Geneva-based watchmaking engineer to draw up blueprints for two new Citizen calibres.[191] At the end of the 1930s, despite the war, Citizen was producing nearly 250,000 watches a year and had become the main challenger of the firm Hattori-Seiko, a role that expanded considerably after 1945.[192] This was a model case of technology transfer via *chablonnage*, with a move from a watch assembly workshop owned by a Swiss importer to a watch manufacturing plant in the hands of local industrialists.

3.2 The maintenance of an industrial district structure

The struggle against *chablonnage* and technology transfer was not the only cause of the cartelization of the Swiss watch industry after World War I. It went together with a strong will to protect the industrial district structure. Mechanization and industrialization of production modes in the years 1870–1920 were limited and did not lead to a real industrial concentration as had occurred in the United States and Japan. In 1923, there were 972 companies involved in watchmaking, employing on average only 35 persons.[193] There was a shared will to maintain this particular structure for three reasons.

The first one was linked to the problem of companies' governance. At the beginning of the 1920s, the overwhelming majority of companies were family firms. For their owners, they were as well the main patrimony, over which they wanted to keep control, as a source of income. This attachment to the individual company explains why the firms' owners favored corporatist measures against the drop of prices due to severe competition, especially among parts makers. Rather than supporting the creation of big industrial companies financed by external capital, in which they had everything to lose, watchmakers preferred the adoption of cartel-like agreements, based on the principle of the generalization of minimum sale prices which would allow each workshop to survive. The experience of the 1906–1909 convention in the sector of gold watches showed it was possible to control the market, and this was the principle on which the cartelization policy of the 1920s and 1930s was based.

The second reason for maintaining the industrial structure was also of an economic nature, but it was not related to firm management. The aim was to maintain a production mode whose flexibility enabled production of a wide range of products, which varied in shape, functions and prices. This plethora of designs was, at the beginning of the 20th century, an essential characteristic of the Swiss watch industry. Industrial concentration would lead to product standardization, which could be beneficial in terms of costs, but it would restrict the diversity of products. This kind of reflection was uncommon at the firm level, where price competition dominated. However, some big manufactures did adopt such strategies. Hélène Pasquier has shown the plethoric supply strategy had been adopted by Omega, which, in the middle of the 1920s, produced luxury jewel watches, sport chronographs and cheap watches. In a catalog published in 1925, the management of this company explained their policy: *"The dominant idea of the managers of this huge organization [Omega] is to make it easy for everyone to acquire a valuable watch whose exterior corresponds to the taste of each country and each customer. It will satisfy the refined amateur, as well as the busy worker, the modest young woman on whom the simple metal or silver wristwatch will smile and the grande dame who will look with complacency at one of these pieces of art gleaming on her wrist."*[194] This thought was also omnipresent within the employers' associations which were defending the common interests of the industry. In 1912, the Société des fabricants d'horlogerie de La Chaux-de-Fonds (Association of Watchmakers of La Chaux-de-Fonds), which had 190 members, published a pamphlet in which it expressed its motto: *"To answer the needs of all countries, all demands, all tastes and all purses."*[195]

Finally, we should mention the social and political factors which played a considerable role in the set up of the watchmaking cartel. The Swiss watch industry was concentrated in the Jura Mountains, outside the big cities, in a region which had few industries other than watchmaking. Industrial concentration, like the production relocation in other countries, would have had a disastrous impact on employment in this region. So, several socialist deputies intervened at the National Council (federal parliament) in the years 1920–1931 to ask for a ban on *chablonnage*, in order to protect the workers. They were supported by the population of the Jura Mountains, which handed a petition signed by some 56,000 persons to the federal authorities in July 1931. Asserting that *"chablonnage favors the manufacture of watches abroad and constitutes a serious threat for the future*"[196], it asked for the intervention of the State in order to put an end to *chablonnage*. For the authorities, the issue was not only to keep employment in Switzerland but also, more largely, to maintain a decentralized production system which kept the social order. Obsessed, after the 1918 general strike, by their fear of trade unionism and communism which would be favored by an industrial concentration within urban centers, the Swiss political elites of the years 1920–1950 wanted to keep workers integrated into local communities, and for this reason supported the maintenance of industrial activity in rural regions.

3.3 The setting up of the cartel

The struggle against *chablonnage* and industrial relocation, and the desire to maintain the structure of the watchmaking sector, led to the reorganization of the industry in the form of a cartel during the 1920s and the 1930s. The setting up of the cartel-like organization was a three-stage process: the adoption of *watchmaking agreements* (1928), the creation of ASUAG (1931) and the legal intervention of the State (1934).

The adoption of watchmaking agreements (1928)

The Chambre suisse d'horlogerie (CSH) played a key role in the setting up of the new structures. As the main meeting point for the watchmakers throughout the country, in January 1923 it called together the representatives of the regional associations to talk about the measures needed for reorganizing the industry. From these discussions emerged the idea of the grouping realized during the 1920s and of the so-called *conventions horlogères* or watchmaking agreements.[197] First, watchmakers were looking for an internal solution by the pooling of the various firms within three groups according to their activity: the watchmakers gathered within the Fédération suisse des associations des fabricants d'horlogerie, (FH, 1924), the main ébauches makers merged into the stock company Ébauches SA (1926), while the other subcontractors came together in the Union des branches annexes de l'horlogerie (Union of watch part makers, UBAH, 1927).

The first to group together were the finished watch producers (*établisseurs* and manufactures). In 1924 they founded the Fédération suisse des associations des fabricants d'horlogerie, (FH), which gathered together the various existing regional organizations. The factory owners of the canton of Bern, where 44% of the employment in the industry was based in 1920, were its spearhead. Grouped together within the Association cantonale bernoise des fabricants d'horlogerie (ACBFH) since 1916, they were very active, especially in their collective negotiation with the workers' union. Besides, very early on this association became conscious of the necessity to bargain fixed prices with parts makers, whose representatives they welcomed into their committee during the first years. This association played a key role in the creation of the FH in 1924. Established in the city of Biel, the FH was successively managed by two former general secretaries of the ACBFH, who left their office to become directors of the FH; they were the lawyers Frédéric-Louis Colomb and Lucien Clerc. In all, the FH consisted of six regional associations (Bern, Le Locle, Fleurier, La Chaux-de-Fonds, Solothurn, and Geneva-Vaud). Finally, one should mention the fact that the roskopf watch makers, specializing in the production of simple and cheap watches, did not enter the FH and stayed out of the conventional system for a while. They did not join together until 1939, within the Association d'industriels suisses de la montre roskopf (Association of Roskopf Watch Swiss Industrialists), also established at Biel. It essentially grouped entrepreneurs from the new watchmaking areas (Solothurn, Basel and Biel) and adopted its own cartel-like agreement, notably concerning minimum prices.[198]

As for the ébauches makers, they gathered within a holding company, Ébauches SA, with its headquarters at Neuchâtel (1926). The three founding firms of this company were the most important Swiss ébauches makers (Fontainemelon SA, A. Schild SA and Ad. Michel SA). They received the financial support of banks (Swiss Bank Corporation, Banque populaire suisse, and the cantonal banks of Bern and Neuchâtel), which appears to have been essential to the rationalization policy. Indeed, until the creation of the ASUAG in 1931, Ébauches SA purchased 31 factories and signed *"friendship agreements"*[199] with eight small ébauches makers. In 1931, it controlled about 90% of the domestic production of ébauches. Unlike the FH and the UBAH, Ébauches SA was not an association but rather an enterprise. It was the only real industrial concentration made in the Swiss watch industry during the 1920s. Its oligopoly position made it a key actor in the business, as it was supplying watchmakers with movements.

Finally, the parts makers grouped together in 1927 within a central association, based on a sectorial organization. The Union des Branches Annexes de l'Horlogerie (UBAH) brought together some fifteen sectorial associations (cases, dials, jewels, springs, etc.).[200] It was based at La Chaux-de-Fonds and at first presided over by César Schild, an industrialist at Grenchen and representative of Ébauches SA; and then, from 1928, successively by the main entrepreneurs of case making and assortments (balances, balance springs, escapements, etc.). The policy of the UBAH at its foundation was that of the subcontractors opposed to industrial concentration. Defending a watchmaking policy based on the principle of minimum prices of parts, it was an essential actor in the cartel and its most fervent supporter until its dissolution in the 1960s. At its foundation, it only grouped together a limited number of makers involved in the production of assortments and *habillage* (the external parts; dials, hands, metal and silver cases). Yet, the success encountered by these makers soon made other makers keen on joining the UBAH: gold case (1932), watch glass (1937), jewel (1941) and pinion (1946) makers joined the central association one after the other, and all the parts makers were members of the UBAH after World War II.

In 1928, these three groups adopted a set of conventions known under the name of "watchmaking agreements" and which essentially concerned three points. First, the various enterprises in these groups engaged to do business only with other enterprises who were members of these groups,

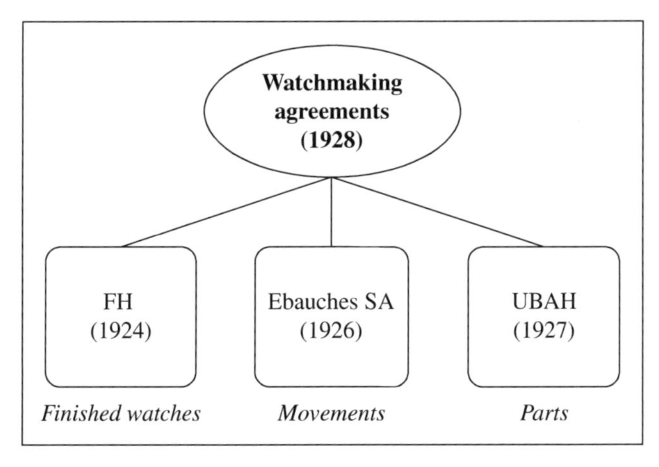

Figure 5: The watchmaking cartel in 1928

Source: Designed by the author.

and to respect official minimum prices. It means that the assembly-makers (*établisseurs*) had to get their supplies of parts from Ebauches SA and the members of the UBAH. From then on, Ebauches SA benefited from a strong position within the industry, with a quasi-monopole. As for the watch companies which produced their own movements in-house, legally defined as "manufactures", they were forbidden to sell movement-blanks (*ébauches*) and parts to other companies. Second, from 1931, these firms accepted not to export any *chablons* (disassembled watches) to anywhere in the world but France and Germany (the "*chablonnage* convention"). And third, they decided to refuse to admit companies founded after 1929 as new members. These agreements had the objective of exerting control on watchmaking through the stabilization of parts' prices, preventing the establishment of new companies and preventing industrial relocation abroad. They were renewed in 1931, and then every five years, and they stayed in force until 1961.

The control and application of these agreements was carried out by a new organization, the Fiduciaire horlogère suisse Fidhor (Swiss Watch Industry Fiduciary), set up with the collaboration of the banks. It undertook investigations into companies at the demand of the *Délégations réunies*, the executive body overseeing the agreements, which was made up of representatives of the FH, the UBAH and Ébauches SA. The employees of Fidhor had access to the accounting of the members of these three groups: they controlled salaries, invoice prices and the volume of production. After 1934, Fidhor also took charge of the application of federal policy in the industry.

Yet the system of watchmaking agreements had a weak point: dissidence. The agreements were private contracts and it was not possible to force enterprises to sign them, and so prevent them from going on with their industrial activities as they wanted. Several small watchmakers, usually *établisseurs* and exporters of disassembled watches, refused to enter the system, especially in the cantons of Bern, Solothurn and Basel. Some of them even grouped together within a so-called independent association in 1930. It essentially gathered together small entrepreneurs and in 1930 they represented only about 5% of all the firms in the business.[201] They were rejecting the growing takeovers by big business in the watch industry. Also, *chablonnage* was still a widespread activity, discretely carried out by many companies, a trend reinforced by the crisis of the 1930s which led to many difficulties. According to Christophe Koller, nearly half of the watchmakers in the canton of Bern were dissidents.[202] The companies which were conspicuously engaged in *chablonnage* were subjected to pressure from the banks and watchmaking associations, which pushed them to join the watchmaking agreements system, but in vain. In March 1931, nine of these small firms gathered together in a Groupement des fabriques d'ébauches suisses (Swiss ébauches factories group) in order to defend their interests.[203] Dissidence was also present among parts makers. In case making for example, some gold case makers of La Chaux-de-Fonds decided in March 1932 to leave the Société suisse des Fabricants de boîtes en or (Swiss gold case makers association) and to group together within a cooperative, Cartelor, with the aim of rationalizing production between them and sharing foreign markets. They were opposed to cartelization and to fixed minimum prices, arguing it was supporting *"poorly organized factories, which can survive only thanks to the tariffs established to give benefits to the more outdated enterprises."*[204]

Setting up a trust: the creation of the ASUAG (1931)

The watchmaking agreements imposed a quota (1928) and then banned *chablonnage*, except for exports to Germany (1931). However, the lack of legal control enabled dissident firms to go on exporting movements and parts around the world. In a message to the Federal Assembly (federal parliament), the government declared in 1931 that *"the partial failure of this reorganization and of the 1928 agreements comes, among other things, from the fact that Ébauches SA does not include all the ébauches factories and that control does not spread to the main parts makers."*[205]

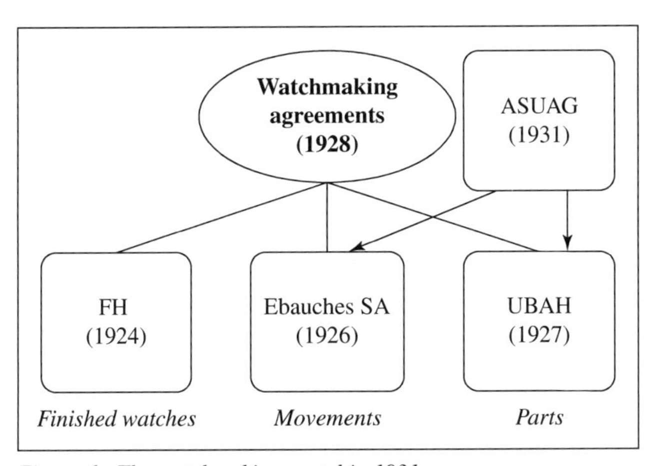

Figure 6: The watchmaking cartel in 1931
Source: Designed by the author.

The issue was then to reinforce the concentration of production for ébauches and assortments (escapements, balance springs and balances). Through the control of production and commercialization of movements, the watchmaker elites hoped to manage the entire watchmaking market.

To cure these problems, Swiss big business intervened in 1931 by creating a super-holding, the Allgemeine Schweizerische Uhrenindustrie AG (ASUAG, Swiss General Watch Industry Co.).[206] The shareholders of ASUAG, whose capital amounted to 10,006,000 francs, were watchmaking industrialists (5 million francs), banks (5 million francs) and the Swiss federal State (6,000 francs in shares, against aid of 6 million francs paid in cash). Furthermore, the State granted a loan without interest of 7.5 million francs and banks supplied several loans for an amount of 15.5 million francs; this total amount of 23 million francs was refunded by the beginning of the 1940s. The financial commitment of the Confederation in a private company was justified by the government by the fact that ASUAG *"is not an ordinary shareholding company with a purely lucrative aim. Its task is to save the interests of the watch industry as a whole."*[207] The Board of Directors included thirty members, among whom five were appointed by the government. Finally, the federal minister in charge of the Department of Public Economy was automatically a member of the executive committee, which shows the importance the State gave to this company.

The objective of the ASUAG was to strengthen the watchmaking agreements system through large-scale industrial concentration. It bought up tens of small dissident ébauches companies, as well as companies involved in the production of special parts. It was thus organized as a super-holding, which controlled four holding companies: Ébauches SA (ébauches), the Société des fabriques de spiraux réunies SA (balance springs), Les Fabriques d'assortiments réunies SA (escapements), and Les Fabriques de balanciers réunies SA (balances), the last three companies being members of the UBAH. The strategy adopted by ASUAG in its concentration policy consisted in buying the majority of the capital in these four holdings, in order to control them, but to allow minority participation to the former owners of these firms, among whom several continued their careers in the watch business until the 1960s. ASUAG followed the way of an industrial concentration which did not clash with the interests of family capitalism, a strategy which explains the success of ASUAG. In the field of ébauches and balance springs, the capital brought by ASUAG enabled it to continue the concentration movement which had begun earlier, while in escapements

Table 13: Shareholders of the ASUAG, 1931

Shareholders	Amount (Swiss francs)
Banks group	*5,000,000.-*
Banque cantonale de Neuchâtel	1,300,000.-
Banque cantonale de Berne	1,250,000.-
Banque populaire suisse (BPS)	1,250,000.-
Swiss Bank Corporation	600,000.-
Federal Bank SA	300,000.-
Union de banques suisses (UBS)	200,000.-
Commercial Bank of Solothurn	100,000.-
Industry group	*5,000,000.-*
FH	1,429,000.-
UBAH	714,000.-
Ébauches SA	357,000.-
Watchmakers	2,500,000.-
Federal State	*6,000.-*
Total	10,006,000.-

Source: *Société générale de l'horlogerie suisse SA. ASUAG. Historique publié à l'occasion de son vingt-cinquième anniversaire, 1931–1956*, Bienne : Arts graphiques SA, 1956, pp. 40–41.

and balances, the creation of ASUAG was the occasion to rationalize. In 1931–1932, Ébauches SA took over seven independent firms, gave financial compensation to seven others for stopping production and bought the equipment used for producing ébauches from ten other firms.[208] The production of balance springs was also concentrated. In 1895, the company La Société des fabriques de spiraux réunies SA (SR) was founded, which merged five big makers. In 1931, when ASUAG bought up SR, there were only two rival firms: the Société suisse des spiraux, in Geneva, and a workshop based at Saint-Imier and belonging to assortments makers. Both were taken over by SR, in 1931 and 1937 respectively.[209] In the cases of assortments and balances, the setting up of ASUAG enabled concentration. The Fabrique des assortiments réunis SA (FAR) was founded in 1932. From then until 1943 it bought up 29 small companies, among which the overwhelming majority were closed down: at the end only eight plants remained, managed by their former owners.[210] As for the Fabriques de balanciers réunies SA, it was also a holding company created in 1932 which grouped together 19 Swiss and foreign workshops.[211]

Talk by the president of the Board of Directors of the ASUAG, 14 November 1942

"When taking over the majority of the shares in the ébauche, assortment, balance spring and balance factories, the aim of ASUAG was to secure a dominant influence in the management of these firms, in order to make them conform to the policy decided by the watchmaking organizations. And the first objective of this policy was to prevent our industry from being transplanted abroad, a very real danger some twenty years ago. That is why we are not a financial company in the common sense. Our aim is to pursue an industrial policy, like many other institutions do under the usual form of an association."

Source: Talk given at a general assembly of shareholders, quoted in *Feuille fédérale*, 1950, p. 75.

The foundation of ASUAG was an essential step in the reorganization of the Swiss watch industry. It embodied the will to rationalize and control the functioning of the industrial district. The objective was not to reconsider the flexibility of the production system. The existence of hundreds of independent watchmakers was the wealth of the Swiss watch industry and enabled it to dominate the world market thanks to the policy of supplying a plethora of designs. Through the creation of ASUAG, bankers and industrialists set up an industrial concentration in the field of watch movements. This strategy had two main objectives. The first one was to control technology transfer and the maintenance of employment in Switzerland, through putting an end to *chablonnage*. The second one was to decrease the production cost of movements. The rationalization of production led to a relative standardization of movements. This policy was also followed by independent manufactures. At Longines for example, new basic calibers were designed in 1928, and then produced in several varieties (size, thickness, kind, etc.), which allowed it to control production costs without reducing the variety of supply.[212]

Because it necessitated massive financial investments, the concentration policy of ASUAG corresponds with the direct intervention of banks into watch business. Of course, they were not newcomers but ASUAG was an opportunity to engage directly in watch companies, as the Banque cantonale de Neuchâtel had been doing at Zénith since 1911. Bankers did not content themselves with a passive role. They had representatives on the boards of directors of both ASUAG and Ébauches SA, where they encouraged a rationalized management. The composition of the Board of Directors of Ébauches SA in 1951 emphasizes their importance (see Table 14). The managers of this firm came from both the ébauches factories themselves and from the banks which invested in the company.

Thus, the position of ASUAG was very particular within the Swiss watchmaking cartel in the sense it was in fact a real trust; that is a company with a monopoly. Except for manufactures that produced their own movements – but got supplies of parts from the subsidiaries of ASUAG – all the Swiss watchmakers had to buy their supplies, ébauches and some special parts (balances, balance springs and assortments) from ASUAG.

Table 14: Board of Directors of Ebauches SA, 1951

Name	Function
Paul Renggli, president	President of the Board of Directors of the ASUAG
Théophile Bringolf	Director of the Banque cantonale de Neuchâtel
Sydney de Coulon	Managing director of Ébauches SA
Fritz Hinderling	Managing director of the Banque populaire suisse (BPS)
Philippe Jéquier	Director of the Fabrique d'ébauches de Fleurier SA
Robert Kaufmann	Watchmaker
Maurice Robert	President of the Board of Directors of the ébauches factory Fabrique d'horlogerie de Fontainemelon SA
Otto Rüfenacht	Member of the Board of Directors of the ébauches factory Felsa AG
Ernst Scherz	Former director of the Banque cantonale de Berne
Adolf Schild	President of the Board of Directors of the ébauches factory A. Schild AG
Rudolf Schild-Comtesse	Director of the ébauches factory Eta AG
Hans Soldan	Director of the Banque cantonale de Berne
Charles Türler	Managing director of the Swiss Bank Corporation
Maurice Vaucher	Vice-president of the ASUAG
Gottlieb Vogt	Member of the Board of Directors of the ébauches factory A. Schild AG

Source: *Les Ebauches: Festschrift zur Feier des fünfundzwanzigjährigen Bestehens der Ébauches-A.G.*, s.l., s.n., 1951, p. 209.

This control of the market for the *watch motor* enabled the avoidance of *chablonnage* and industrial transplantation abroad; but it especially gave ASUAG a dominant position within the industry, with the objective of securing and beneficially applying the massive investments of the banks. Until the quartz revolution, which would enable Swiss and foreign makers to get cheap movements elsewhere, ASUAG was the key actor in the Swiss watch industry.

The legal intervention of the State (1934)

The industrial concentration realized by ASUAG was however not all-comprehensive and several dissident companies emerged again, resulting in the continuation of *chablonnage*. An internal report written by ASUAG focused on this problem in autumn 1933.[213] There were then 22 small firms actively involved, among which nine were ébauches makers, six assortment and seven balance spring makers. As for the watchmakers, there were only ten which had not signed the agreements (8 *établisseurs* and 2 manufactures). Their overall number was not important but sufficient to challenge the functioning of the system.

In the end, only the legal intervention of the federal State could put an end to dissidence. On principle the watchmaking big businesses opposed such a policy, but moderated their reluctance with the 1930s crisis, during which watch exports dropped from 20.8 items in 1929 to 8.2 million in 1932. They accepted the involvement of the State, which could *"close the circle"*[214], as said the federal councilor Schulthess, head of the Department of Public Economy. The Confederation passed three federal decrees in 1934–1936 which resulted in the legalization of the watchmaking agreements system of 1928. The first one was the decree of 12 March 1934, which required licenses for opening new companies, increasing the workforce (itself subjected to quota for each firm), expansion of factories, diversification into new products, and relocating the headquarters of existing companies. In addition, the export of *chablons*, ébauches and parts were also subjected to official licenses. The objectives of this first decree were to manage the structure of the industrial district and to control the practice of *chablonnage*. The implementation of this policy was done by the Federal Department of Public Economy, with the collaboration of the CSH and Fidhor. The system was strengthened in 1936 with the adoption of two new decrees: the first to legalize the principle of minimum prices set up in the 1928 agreements, to which dissident companies

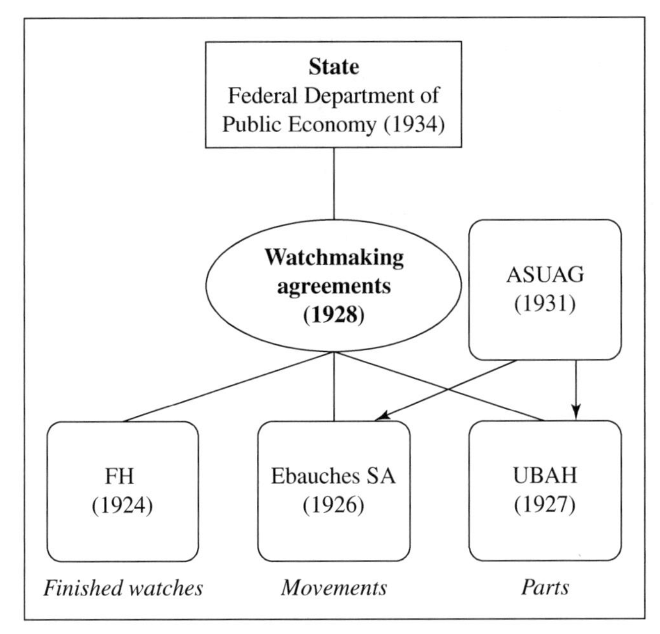

Figure 7: The watchmaking cartel in 1934

Source: Designed by the author.

also had to submit from then on (decree of 13 March 1936); and the second to regulate home work, which was limited to some specific activities (decree of the 9 October 1936). These three decrees were unified and renewed in 1937, and then renewed several times until 1961. They constitute the core of the economy policy which defined the watch industry for several decades, known under the name of the watchmaking statute (*Statut horloger*).

The massive intervention of the State to support the watch industry, in both financial (ASUAG) and legal (decrees) aspects, was surprising. Traditionally, the Swiss Confederation rarely intervened in industry, even in the case of sectors coping with difficulties, like the textile industry in the interwar years. The crisis was of course an influential factor, but not sufficient to explain such a strong intervention, which cannot be seen in other sectors.

In addition, the will to maintain employment in Switzerland is not enough, as other industries, like chocolate, massively delocalized their production during the 1920s and 1930s without the government becoming especially concerned about it.[215] The importance of banks seems to be a decisive factor of the State intervention. Through a very large credit policy, they lent a lot to small and medium sized watch firms, and invested in ASUAG during the interwar period. At the same time, they faced severe losses abroad, especially in Germany. Some of them, like the Banque populaire suisse (BPS), saved by the federal State in 1933, faced serious financial difficulties. So banks may have played a key role in favoring the intervention of the State in watchmaking. Through the legalization of the 1928 agreements system and the trust, the federal decrees of 1934–1936 enabled the stabilization of the structure of the watch industry and to secure the financial involvement of banks.

Reinforced by the intervention of the State, the agreements system of 1928 gave birth to a real bureaucracy, whose main centers were the cities of Bern (Federal Department of Public Economy), Biel (ASUAG and FH), La Chaux-de-Fonds (CSH and UBAH) and Neuchâtel (Ébauches SA). Until the beginning of the 1960s, watchmaking companies no longer worked in a liberal environment, but rather within the framework of an interventionist economy. Key elements such as the kind of parts produced (technical characteristics and sale prices), the engagement of workers, the opening of new firms and the export of parts were submitted to licensing by the Federal Department of Public Economy. The triptych of State, watchmaking firms and banks was the basis of the watchmaking cartel for about thirty years. One of the main personalities of this system was without doubt Sydney de Coulon (1889–1976), a man who came from finance and became one of the key leaders of the watch industry under the regime of the cartel.[216] Descended from a former aristocratic family of Neuchâtel, he was the son of a private banker and began his career in this business. After an apprenticeship in the United Kingdom, he worked in a bank at Lausanne and afterwards he became a member of the Board of Directors of the Swiss National Bank. His marriage to a daughter of Paul Robert-Tissot, head of the ébauche factory of Fontainemelon, one of the biggest in Switzerland, led to his involvement in the watch business. He became a director of this firm and from the 1930s participated in the management of the new companies created by the concentration process. Thus, he was notably executive officer of ASUAG (1931–1933), managing director of Ébauches SA (1932–1964) and member of the Board of Directors of the Fabriques d'assortiments réunies (FAR) in the 1950s. Finally, de Coulon began a political career in the 1940s. A member of the Liberal Party, he was a member of the legislative council of the canton of Neuchâtel (1941–1954), and deputy at the National Council (1947–1949) and then at the State Council (1949–1963), the two chambers of the Swiss federal parliament. Through his multiple engagements, Sydney de Coulon appears as a man linking watchmaking, banking and politics. He thus perfectly personifies the watchmaking cartel.

The labor peace agreement

The intervention of the State in the economic and industrial field was finally completed in 1937 by the adoption of an agreement between employers and trade unions. It established the labor peace agreement and is also an essential element of the cartel. The origin of this agreement goes back to the intervention of the federal authorities during the autumn 1936. The devaluation of the Swiss franc in September 1936 threatened to lead to a rise in the cost living and then to social conflicts in industry, so the Federal Council passed a decree stating that *"the Federal Department of Public Economy will arbitrate, without appeal, conflicts on wages which would spread to more than one canton and which could not be resolved by involved parties. For this purpose, the establishment of a wages joint commission is envisaged."*[217] The pressure of the political authorities on employers' associations and trade unions made them negotiate and adopt an agreement on the main questions related to working conditions in the watch industry.

Yet, the intervention of the State was not the sole reason why a labor peace agreement was adopted in the watch industry.[218] In the middle of the 1930s, the trade union dramatically shifted its political position which helped the negotiations with the employers. Thus, in 1934 the FOMH gave up Marxist references to class struggle and from then on favored the improvement of the working conditions of its members through dialogue. In 1926, the trade unionist Achille Grospierre, a member of the National Council, had already called for the intervention of the State to put an end to *chablonnage*, arguing it would help to save employment in Switzerland.[219] Moreover, at the beginning of the 1930s, the FOMH was involved in the setting up of the cartel's bodies. The Federal Council gave, for example, two of its five places on the ASUAG Board of Directors to representatives of the FOMH.

Yet, at first, this new direction of the FOMH was not welcomed by some workers, who wanted to continue the confrontational approach, as well as by some watchmakers who did not believe it was a true change. It was only after the launching of a strike in March 1937 in a dial factory at Biel, for changes in wages, that the employers, threatened by the spread of the conflict to other companies, engaged discussions with the FOMH and accepted to negotiate about wages. These discussions between the trade union and the employers' associations developed the framework which led to the signing of the work collective convention of 15 May 1937, through which the partners engaged themselves to implement an *"absolute social peace regime"*[220]: the FOMH abandoned the strike and employers ended the lock-out.

This first work collective convention led to the creation, in January 1938, of a new body, called the Collective convention between watchmaking employer's associations (Convention collective entre les associations patronales horlogères, Convention patronale since 1966), with the aim to gather all the employers' associations into a single organization which could negotiate with the trade unions. From 1937 on, the main improvements in working conditions resulted from negotiations within this framework. It led, for example, to the introduction of a family allowance (1942), birth allowance (1946), contributions to health insurance (1947), and the progressive decrease of working hours (see Table 15).

Table 15: Weekly working hours in watch companies who were members of the collective convention

Year	Weekly working hours
1916	58.5 h.
1919	52.5 h.
1919	48 h.
1923	52 h.
1937	48 h.
1957	47 h.
1958	46 h.
1960	45 h.
1963	44 h.
1977	43 h.
1979	42 h.
1985	41 h.
1987	40 h.

Source: Beck, Renatus (dir.), *Voies multiples, but unique. Regard sur le syndicat FTMH 1970–2000*, Lausanne, Payot, 2004, pp. 140–144.

3.4 The consequences of the cartel

The framework of the cartel within the Swiss watch industry developed for about three decades, and had as its main consequence the maintenance of the decentralized structure of this industry. Yet, it did not prevent new rival foreign companies from emerging and establishing themselves on the world market.

The maintenance of the structures

The federal census of firms taken in 1955 showed the persistence of the atomization of production among numerous small enterprises. In that year, there were in Switzerland a total of 2806 companies involved in watchmaking: 2241 (79%) employed 20 or less persons; 312 (11%) between 21 and 50 persons; 127 (5%) between 51 and 100 persons; and only 126 had more than 100 employees.[221] However, the concentration of firms was not completely absent but limited to some sectors, in particular movements and their parts (ébauches, balances, assortments and balance-springs). These were actually activities controlled by holding companies which belonged to ASUAG. Except for them, the structure of the watch industry was maintained.

This conservatism resulted from a policy developed and applied by the federal authorities, in agreement with the leading watchmakers. The federal decree of 1934 required that a range of matters related to the organization of the industry be officially authorized. The opening, enlargement, geographical movement and change of corporate legal status of the firms depended on a license given by the federal Department of Public Economy. It was the same for increasing the workforce, as each enterprise was subjected to a quota of workers. Finally, changing industrial activities also depended on an official license. For example, a silver case maker needed a license to shift to metal or gold cases. This very strict control of company activities enabled the State and the watchmaking leaders to decide the evolution of the means of production.

The application of this policy reveals a clear-cut will to favor the maintenance of existing structures. While the industry was in a high growth trend, characterized by the increase of exports from 15.2 million

Table 16: Requested and granted licenses by the Department of Public Economy, 1937–1959

	Requested	Granted	Granted as a %
Opening of a new company	4464	1160	25.9
Enlargement of an existing company	1965	1722	87.6
Shift to a new activity	1373	541	39.4

Source: *Feuille fédérale*, 1950, pp. 116–117 and *Feuille fédérale*, 1960, pp. 1512–1513.

pieces in 1935 to 24.2 million in 1950 and 40.9 million in 1960, the State tried to limit, as much as possible, the dispersion of the industrial activities during the years 1937–1959 (see Table 16). It was reluctant to allow the opening of new companies (25.9% of requests granted) and restricted shifts to new activities (39.4%). However, the enlargement of existing firms was usually approved (87.6%). It was the same for geographical movement of firms (92.6% of requests granted in the years 1937–1950).[222]

In allowing the maintenance of many small enterprises throughout the Jura Mountains, the cartel adopted the dominant ideology which was fearful of the political and social effects of industrial concentration in urban centers. The fear of communism and of the rise of a protesting trade-union was omnipresent in the watchmaking policy of the federal authorities. Addressing the Federal Assembly in 1950, the government expressed its view, arguing that watchmaking *"includes numerous artisanal-like small enterprises; their owners belong to the middle class, so necessary for the social and political balance of the country. These small firms disseminated in villages and towns give to many persons the possibility of earning a living without having to leave their locality. At the same time they ensure the community has an appreciable fiscal resource. [...] It is important to protect the small and medium sized enterprises, on the same level as the big ones, and to maintain the decentralization of watchmaking, which may be easier to achieve than in other industries."*[223] Indeed, the control of prices and the quasi absence of competition between subcontractors ensured the viability of many workshops. Watchmaking was an industrial sector which experienced high profits. The dividends paid by watchmaking companies to their shareholders during the years 1939–1956 averaged 14.0% against 7.8% for industry as a whole (Table 17).

Table 17: Dividends paid by the Swiss stockholding companies, as a %, 1939–1956

	Watchmaking	Industry as a whole
1939	8.17	6.08
1940	8.2	6.97
1941	8.71	7.56
1942	8.01	7.02
1943	10.22	6.82
1944	9.01	6.54
1945	12.66	6.73
1946	16.25	7.26
1947	18.37	9.83
1948	28.98	8.29
1949	11.14	7.29
1950	13.51	8.14
1951	18.02	8.24
1952	14.48	8.3
1953	18.61	8.47
1954	14.61	8.75
1955	15.73	8.62
1956	16.72	8.62

Source: Rieben, Henri, Urech, Madeleine and Iffland, Charles, *L'horlogerie et l'Europe*, Neuchâtel: La Bacconière, 1959, p. 193.

However, the maintenance of the structures of the industrial district had a negative impact on the organization of the production system, especially for manufacturers. As both the sale of parts outside the company and the merger of other firms were strictly controlled, and usually refused, the manufacturers encountered financial difficulties due to the necessity to develop new products, on the one hand, and the impossibility to rationalize production, on the other hand. This lack of rationalization is the main reason why most of them closed down production facilities and were taken over after 1960, when markets became more competitive.

The creation of the Société suisse pour l'industrie horlogère SA (SSIH)

Besides ASUAG, a second industrial group was founded during the interwar years and became one of the major actors in the Swiss watch industry. This was the Société Suisse pour l'Industrie Horlogère SA (Swiss Company for the Watch Industry, SSIH), founded in 1930 as the result of the merger of the two companies Omega and Tissot. Until the beginning of the 1960s, the SSIH was the main group which gathered together makers of finished watches.

A third group, the Société Anonyme en Participations Industrielles et Commerciales (SAPIC), was created in 1927 by the manufacture LeCoultre & Cie SA and its French partner, the Etablissements Jaeger. Later it acquired shares in its customers, such as Vacheron & Constantin (1938) and Audemars Piguet (1948).[224]

The financial difficulties encountered by the watch companies during the interwar period made it necessary to rationalize production. In the middle of the 1920s, the main manufacturers of the country, namely Omega, Longines, Tissot, Vacheron & Constantin, IWC and Zénith, engaged in discussions and negotiations, the context which led to the SSIH. Tissot and Omega were both facing severe conditions. Tissot had to deal with the closure of its main outlet, Russia. Its sales dropped from 27,000 watches in 1920 to 11,000 in 1922; that is, a decrease of 60%. And Omega also experienced a decrease, with its sales going down from 246,000 pieces in 1920 to 78,000 in 1921, a fall of 75%. The financial situation of the firm was very severe: debts amounted to 2.8 million francs in 1924. These difficulties led both companies to get closer and to adopt a rationalization strategy. In 1924 they signed an agreement by which they engaged to collaborate on the commercial (marketing) and technical levels (Tissot producing some calibers for Omega). Finally, both firms decided to share the market between precision watches (Omega) and standard quality watches (Tissot). The SSIH was formally created in 1930 with the aim of strengthening ties between the two companies.

Afterwards, the SSIH enlarged its production capacity through the merger of other firms. In 1932 it bought Lemania Watch Co Lugrin SA, and then, in 1953, the company Marc Favre, specialized in women's watches. In addition, the production capacity of Tissot was made available to Omega, which faced its highest growth after the war; thus, in 1955 Tissot manufactured 60,000 movements for Omega.[225] Despite their merger within a group, the various companies taken over by SSIH continued their own development. The strategy adopted by the management of SSIH was to increase the production capacity of Omega through the purchase of other firms. At the occasion of the take-over of the company Marc Favre, the Board of Directors of SSIH declared that *"with this additional production, Omega will be able to satisfy the ever growing needs of its customers."*[226] Anyway, the rationalization process could not be completely achieved due to the reluctance of the cartels to allow a too strong industrial concentration.

The failure of the struggle against chablonnage and the emergence of new watchmaking nations

Even if it succeeded in controlling the evolution of the structure of the industry, cartelization appears to have failed in its attempt to prevent new rivals emerging through the banning of technology transfer. At first, the cartel never really put an end to *chablonnage*. The export of movements did not stop and was even growing under the cartel regime. Their volume went from 3.9 million pieces in 1935 to 7.4 million in 1950 and 10.8 million in 1960. In relative numbers, these movements represent 28.8% of the volume of the export of watches in the 1950s.

The United States was still the main country to which *chablonnage* was directed. The growth of the number of American watches equipped with foreign movements shows the ineffectiveness of the measures adopted in Switzerland (Figure 8). The volume of these watches amounted

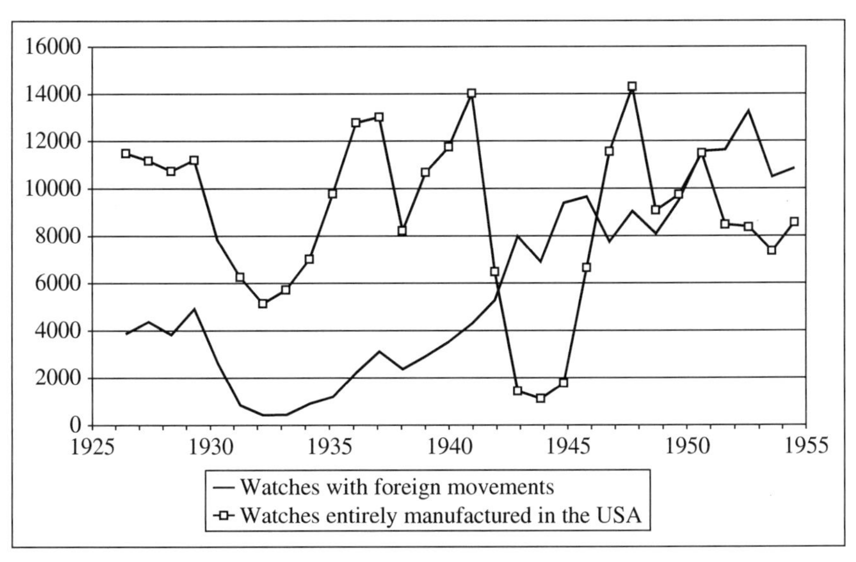

Figure 8: USA watchmaking production, thousands of units, 1926–1955

Source: Landes, David S., *Revolution in time, clocks and the making of the modern world*, Cambridge: Harvard University Press, 1983, p. 385.

on average to 4.2 million pieces per year during the period 1926–1929, against 11.2 million of watches completely manufactured in the United States. After a decrease due to the great depression (less than 500,000 watches with foreign movements in 1932 and 1933, and a national production which dropped to less than 6 million pieces), the number of watches equipped with foreign movements entered a growth trend for about twenty years: their number reached 1.2 million of pieces in 1935, 9.4 million in 1945 and 10.9 million in 1955. After 1951 the production of this kind of watch was even higher than the American domestic production.[227] It was mainly the consequence of the loss of competitiveness of American watch companies due to the concentration on munitions production, such as bomb fuses, during World War II.

The continuation of *chablonnage* can be explained by two reasons. First, the export of movements was not completely banned by the federal decree of 1934, but rather regulated. The sale of ébauches and parts to former customers was tolerated. This was especially the case of American firms, such as Bulova and Gruen, which possessed subsidiaries in Switzerland. These companies assembled and cased Swiss made movements, as they had been doing since the beginning of the 20th century. Besides, the question of the export of parts from Swiss branches to American headquarters was a difficult problem for the cartels, as this trade occurred within a single company, and was not properly defined by the regulations of the cartel adopted in the 1930s. Second, even if the concentration strategy of ASUAG was efficient in the sector of movement parts makers, this company did not control the activities of the so-called "manufacturers", that is, firms which were producing both movements – in principle for their own needs – and finished watches. Yet some manufactures participated in the export of disassembled movements to some of their agents, notably in the American market upon which they depended. The ambiguous role played by the ASUAG, the UBAH and the FH must be highlighted here: they signed some agreements with Gruen Watch (1943), Benrus Watch (1945) and Bulova Watch (1948) to authorize the export and the production in America of some parts.

Thus, in the context of a constantly expanding market until 1970, the existence of the cartel did not prevent either technology transfer or the emergence of new watchmaking nations on the world market after 1945 (see Table 18). The appearance of new technologies (pin lever, electrical and tuning fork watches) outside Switzerland also helped this phenomenon. Making possible the mass production of precise and cheap products, these technologies favored, before the quartz revolution, the emergence of newcomers. Thus, the Swiss watch industry

Table 18: World watch and movement production, 1945–1970

	1945	1950	1955	1960	1965	1970
Total (1,000 pieces)	21,567	47,723	73,557	99,385	122,800	176,746
Switzerland, %	87.2	52.4	47.2	42.5	44.6	41.6
USA, %	8.3	20.5	11.5	9.6	11.0	11.0
Japan, %	0.2	1.5	3.0	7.2	11.1	13.5
USSR, %	0.5	4.5	11.9	16.6	13.0	12.4
France, %	3.9	6.7	5.0	5.3	5.3	6.2
Germany, %	0.0	7.3	9.5	8.2	5.9	4.6
Others, %	0.0	7.1	11.9	10.6	9.1	10.6

Source: Landes, David S., *Revolution in time, clocks and the making of the modern world*, Cambridge: Harvard University Press, 1983, pp. 386.

coped after 1945 with competition on two levels. First, it was facing, in a majority of cases, industries oriented to their own domestic market (USA, France and the United Kingdom), or a regional market, as the USSR in communist countries. Not export oriented, these industries usually benefited from State protection which restricted the import of foreign watches and thus constituted a barrier to the Swiss watch trade. Second, some countries, especially Germany and Japan, did not limit themselves to their domestic market but exported a growing part of their production onto the world market, where they challenged Swiss watches thanks to cheap and reliable products. In 1958, Germany exported 45.5% of its horology production, mainly consisting of clocks which were not directly in competition with Swiss watchmakers.[228] As for Japan, it was at that time still oriented to the domestic market and exported only a small part of its watch production (0.9% in 1955 and 2.9% in 1960). But it was preparing a large commercial offensive on the world market which led its watch exports to rise to 32% of national production in 1965 and made it a formidable rival for Swiss watchmakers.[229]

The success of the watchmaking nations which established themselves as the main competitors of Switzerland in the 1950s, principally the United States and Japan, was based on the adoption of mass production systems and on the use of international labor. The company U.S. Time Corporation is an excellent example of the organization of American watch companies on a global scale. In 1960, it had an ébauches factory in Besançon (France) and an assembly workshop at Porto-Rico, as well as a subsidiary in the United Kingdom which was realizing about one third of British watch production at the end of the 1950s.[230] The companies Hamilton and Bulova had similar structures.

3.5 New products, new markets

Even if the cartelization of the watch industry led to the locking in of production structures, Swiss made watches changed significantly during the years 1920–1960, the main feature of which was the democratization of their use; the manufacture of Swiss watches had a general trend towards mass production of cheap watches. This mutation was in response to a shift in the world market, which had two distinct phases during this period.

The first phase is one of stagnation, which ran from 1920 to 1936. While Swiss watch exports had a quasi permanent growth from the 1880s to World War I, the interwar period is marked by two important crises (1920–1921 and 1930–1935) whose social impact was particularly pronounced: in the watch industry, the rate of full unemployment rose to 12% in 1933 and 19.4% in 1934, and that of partial unemployment to 10.4% and then 19.5%.[231] The second phase is one of growth. It began in 1936 with the devaluation of Swiss franc and went on nearly without any break until 1974. The decrease in watch exports observed during World War II was not a crisis, as the value was increasing during these years, rising from 196 million francs in 1939

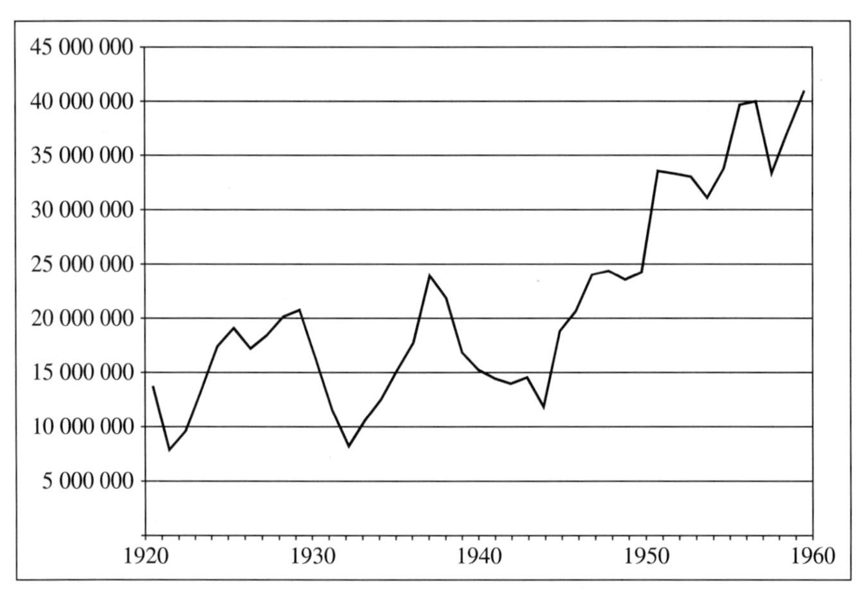

Figure 9: Swiss exports of watches and movements, volume (1,000 pieces), 1920–1960

Source: *Statistique du commerce de la Suisse avec l'étranger*, Berne : Département fédéral des Douanes, 1920–1960.

to 493 million in 1945, with military production allowing a development of business despite the fall in sales of watches.

It was especially during the first phase that the watch industry adapted its products to a changing business environment. The ébauches and parts factories concentration policy made it possible to largely rationalize production of movements. Under the auspices of the Chambre suisse d'horlogerie (CSH), a new body was created at the beginning of the 1920s in order to supervise the standardization of parts: the Normalisation Horlogère Suisse (NHS). Moreover, some manufactures – who produce their own ébauches according to the legislation – also adopted a rationalization policy from the 1910s to the 1930s, characterized by the reduction of the number of their different calibers with the aim of cutting costs.[232] Even the most prestigious manufacturers, such as Longines, launched some simple and cheaper watches which made it possible to be competitive on a market in crisis. This rationalization policy was pursued in enterprises as well as at the level of the whole industry. It allowed the Swiss watch industry to keep its competitive advantage on the world market thanks to a fall in prices. The average value of Swiss watches exported in the years 1920–1935 was

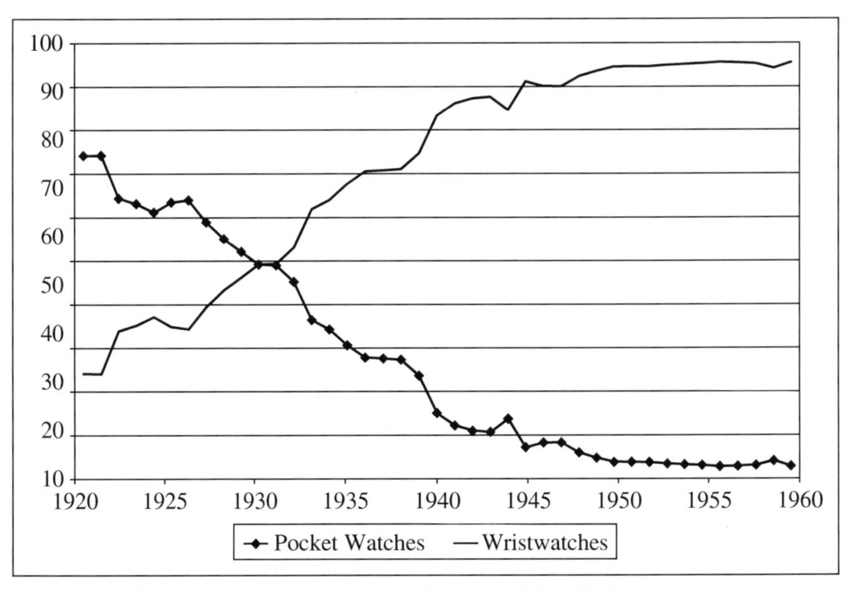

Figure 10: Kinds of Swiss watches exported, as a %, 1920–1960

Source: Ritzmann, Heiner (ed.), *Statistique historique de la Suisse*, Zurich : Chronos, 1996, p. 627.

indeed continuously decreasing: it went down from 23.7 francs in 1920 to 8.2 francs fifteen years later.

This rationalization was accompanied by a substantial mutation of the product itself, with two characteristics. The first one is the shift from pocket watches to wristwatches as the dominant product. Even though it had existed since the end of the 1890s, the wristwatch amounted to only to a quarter of Swiss exports at the beginning of the 1920s. But it took on a growing importance during the interwar period and exceeded pocket watches in 1930. The second characteristic is the decrease in precious metal watches (gold and silver) compared with base metal watches (steel and nickel). While they still amounted to 51.8% of the export volume in 1920, gold and silver watches were only 29.1% in 1930. In 1935, they fell below the 5% mark and did not rise above that until the 1960s. Thus, an important mutation of the product occurred during the interwar years, with the shift from the precious metal pocket watch to the cheap wristwatch. The modification of Swiss watch production during the crisis years set up the foundations for mass production which characterized the high growth period. Low quality roskopf watches became very numerous after 1936 and they represented 28.7% of the volume of exports in the years 1937–1953.[233] However, there was a conflict between the evolution of the product and of markets, which necessitated a more intensive industrial concentration, and the organization of the cartel which prevented achieving the rationalization process. In this conflict lies the origin of the 1970s and 1980s crisis.

Despite the trend towards mass-produced watches, Swiss companies pursued the development of advertising oriented to luxury and precision and they were selling more and more standardized products as top-of-the-range watches. However, some companies involved in developing new products sought an increasingly better precision. From the end of the 19th century, chronographs particularly became an object of competition between several manufactures whose aim was to produce watches able to measure time more and more precisely. These timekeepers were then used in sports competitions, especially in the Olympics, and were soon included within a new advertising strategy.

Longines was one of the first firms to produce pocket chronographs industrially, on the basis of a caliber designed by the watchmaker Alfred Lugrin in 1878.[234] Afterwards, it was developed within the firm and led to a 1/5 second chronograph which was probably used for the Olympics at Athens (1896). Like Longines, the company Heuer, based then at

St-Imier, also produced chronographs from the 1880s. In 1916, this firm launched on the market a 1/100 second chronograph, the Mikrograph, with which Heuer was able to become the official timer at the Olympics during the 1920s (Anvers 1920, Paris 1924, Amsterdam 1928). Finally, as the production of chronographs became a real challenge for advertising and communication, in 1932 Omega decided to enter this field by merging with the small company Lemania-Lugrin SA, founded in 1918 by Alfred Lugrin and specializing in the production of high-precision chronographs.[235] In 1932 Omega was awarded the official timing of the Los Angeles Olympics and held this privilege until 1964, when it was dethroned by Seiko on the occasion of the Tokyo Olympiads. While the rival firms Longines and Heuer did not use their title of official timer of the games in long term advertising strategies, Omega did.

Thus, as the example of chronographs highlights, sports timing and prizes obtained at precision chronometer trials became increasingly important advertising issues. Yet, this marketing strategy was not limited to a few manufactures but rather touched all the watch exporters, who listed in their advertisements the various prizes obtained in chronometer trials. They had increasing recourse to the watch control offices of the different watchmaking schools, and the obtaining of a certificate became a tool for publicity. At the control office of Biel for example, the number of watches tested grew from 350 in 1918 to 1,190 in 1928.[236]

The market outlets also showed a significant transformation during the cartel regime, with two features: the return of the USA as the main market and a general diversification of second rank markets. In the 1900s, the Swiss watch industry had become emancipated from dependency on America thanks to a growth of sales in Europe. Yet, in the interwar years and in the 1950s, the United States again became the main outlet. Even if it was at first *chablonnage* which supported this growth, the setting up of the cartel seems to have had no effect on this trend, the drop observed in the 1930s being the consequence of the economic crisis. Between 1933 and 1945, the United States had a growing importance, going from 7.7% to 49.1% of the value of Swiss watch exports (see Figure 11). This increase can be partly understood as the result of armaments' orders, but it was also the consequence of the activities of American watch companies like Bulova and Gruen, which had subsidiaries in Switzerland that supplied their American factories. The vital importance of the American market led ASUAG to revise its anti-*chablonnage* policy in the case of the United States. After 1945, the relative importance of the American market was

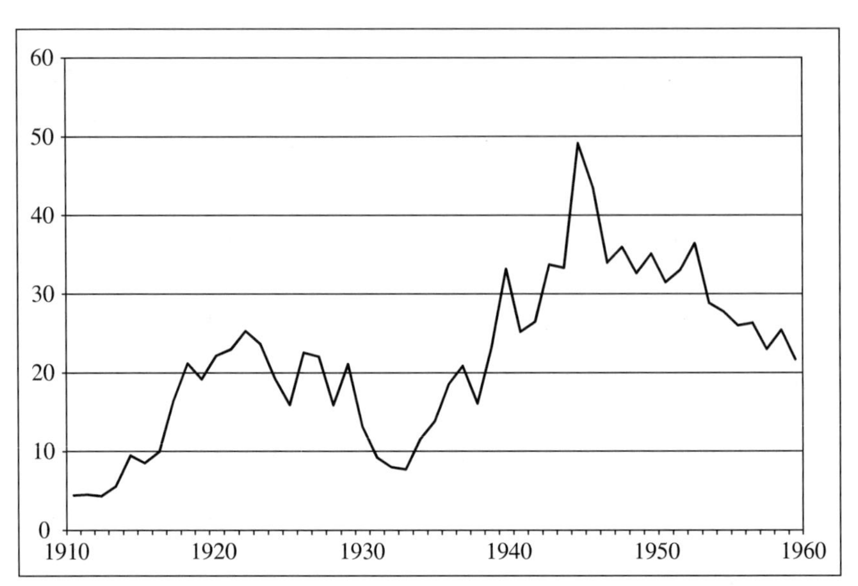

Figure 11: Part of the United States in Swiss watches exports (value as a %), 1910–1960

Source: *Statistique du commerce de la Suisse avec l'étranger*, Berne : Département fédéral des Douanes, 1910–1960.

declining but that is more a consequence of the emergence of new markets than a real stagnation of this outlet. Measured as a volume of pieces, the exports to the United States grew during the 1950s (8.9 million in 1950 and 13.6 million in 1960).

The second feature of Swiss watch exports in the first part of the 20th century is the diversification of outlets. Europe, which amounted to about two thirds of sales in the 1900s, represents only 32.4% of exports in 1960. Other than the United States, the main growth market was Asia (20.5%), even though the Japanese market was not freely opened until 1961. Latin America (18.6%), Africa (5.0%) and Oceania (1.9%) were still secondary markets.

Notes

166 *Feuille fédérale*, 1921, p. 480.

167 Pasquier, Hélène, *La "Recherche et Développement" en horlogerie. Acteurs, stratégies et choix technologiques dans l'Arc jurassien suisse (1900–1970)*, University of Neuchâtel , PhD thesis, 2007, p. 31.

168 *Feuille fédérale*, 1921, p. 482.
169 *Feuille fédérale*, 1922, p. 195.
170 For the practice of *chablonnage*, see Koller, Christophe, *L'industrialisation et l'Etat au pays de l'horlogerie. Contribution à l'histoire économique et sociale d'une région suisse*, Courrendlin : CJE, 2003, pp. 374–380.
171 *Feuille fédérale*, 1931, p. 199.
172 *Société générale de l'horlogerie suisse SA. ASUAG. Historique publié à l'occasion de son vingt-cinquième anniversaire, 1931–1956*, Bienne : Arts graphiques SA, 1956, p. 16.
173 Koller, Christophe, *"De la lime à la machine". L'industrialisation et l'Etat au pays de l'horlogerie. Contribution à l'histoire économique et sociale d'une région suisse*, Courrendlin : CSE, 2003, pp. 379–381.
174 Maeder, Alain, *Gouvernantes et précepteur neuchâtelois dans l'empire russe (1800–1890)*, Neuchâtel : Institut d'histoire, 1993, p. 113.
175 Fallet, Estelle, *Tissot, 150 ans d'histoire*, Le Locle : Tissot SA, 2003, p. 222.
176 *Société générale de l'horlogerie suisse SA. ASUAG. Historique publié à l'occasion de son vingt-cinquième anniversaire, 1931–1956*, Bienne : Arts graphiques SA, 1956, p. 16.
177 Harrold, Michael C., *American Watchmaking. A Technical History of the American Watch Industry, 1850–1930*, Columbia: NAWCC, 1984, pp. 35–36.
178 Koller, Christophe, *"De la lime à la machine". L'industrialisation et l'Etat au pays de l'horlogerie. Contribution à l'histoire économique et sociale d'une région suisse*, Courrendlin : CSE, 2003, p. 439.
179 Bolli, Jean-Jacques, *L'aspect horloger des relations commerciales américano-suisses de 1929 à 1950*, La Chaux-de-Fonds : La Suisse horlogère, 1956, p. 79.
180 Landes, David S., *L'heure qu'il est : les horloges, la mesure du temps et la formation du monde moderne*, Paris : Gallimard, 1988, p. 458.
181 Koller, Christophe, *"De la lime à la machine". L'industrialisation et l'Etat au pays de l'horlogerie. Contribution à l'histoire économique et sociale d'une région suisse*, Courrendlin : CSE, 2003, pp. 419–431.
182 Uchida, Hoshimi, *Tokei sangyou no hattatsu*, Tokyo: Seiko Institute, 1985, 494 p. and Donzé, Pierre-Yves, "Le Japon et l'industrie horlogère suisse. Un cas de transfert de technologie durant les années 1880–1940", *Histoire, Economie et Société*, 2006, pp. 105–125.
183 Makoto, Ichihara, *Yume o utta otoko. Kindai sangyo no paionia. Tenshodo – Ezawa Kingoro*, Tokyo: Ronshobô, 1990, 173 p.
184 Jaquet, Eugène and Chapuis, Alfred, *Histoire et technique de la montre suisse*, Bâle : Urs Graf, 1945, p. 142.
185 Ezawa, Tomikichi, *Nanajûnana ô kaikodan*, Tokyo : Shikaishobô, 1939, p. 112.
186 Musée international d'horlogerie, La Chaux-de-Fonds (MIH), report on *chablonnage* in Japan, 6th of October 1930.
187 MIH, ASUAG's report, 5th of October 1933.
188 Uchida, Hoshimi, *Tokei sangyou no hattatsu*, Tokyo: Seiko Institute, 1985.
189 Chou, Kouken, *Nihon tokeisangyō no hatten to kigyō ka katsudō. Daiichijidaisen izen no Seikosha to jirei tochite*, Kyoto University: master's thesis, 2002, p. 23. This census mentions as well two clockmacking firms employing more workers than

Schmid: Aichitokei (702 workers) and Nagoya Shoji (172 workers). But they do not produce watches.

190 *Shashi*, Tokyo: Citizen Life, vol. 1, 2002, 156 p.

191 AFS, E 7004, 1967/12.

192 *Shashi*, op. cit., p. 11. For comparison, the average annual output of the Swiss firm Longines is less than 200'000 watches during the period 1938–1940.

193 *Feuille fédérale*, 1931, p. 193.

194 Quoted by Pasquier, Hélène, *La "Recherche et Développement" en horlogerie. Acteurs, stratégies et choix technologiques dans l'Arc jurassien suisse (1900–1970)*, University of Neuchâtel , PhD thesis, 2007, p. 56.

195 Bubloz, Gustave, *La Chaux-de-Fonds, métropole de l'industrie horlogère suisse*, La Chaux-de-Fonds : Société des fabricants d'horlogerie de La Chaux-de-Fonds, s.d. [1912], p. 12.

196 *Feuille fédérale*, 1931, p. 206.

197 Koller, Christophe, *"De la lime à la machine". L'industrialisation et l'Etat au pays de l'horlogerie. Contribution à l'histoire économique et sociale d'une région suisse*, Courrendlin : CSE, 2003, pp. 353–353.

198 Koller, Christophe, *"De la lime à la machine". L'industrialisation et l'Etat au pays de l'horlogerie. Contribution à l'histoire économique et sociale d'une région suisse*, Courrendlin : CSE, 2003, p. 355.

199 *Société générale de l'horlogerie suisse SA. ASUAG. Historique publié à l'occasion de son vingt-cinquième anniversaire, 1931–1956*, Bienne : Arts graphiques SA, 1956, p. 59.

200 Wyss, Jean, *La création de l'Union des Branches Annexes de l'Horlogerie (UBAH) et les vingt premières années de son activité (1927–1947)*, La Chaux-de-Fonds : UBAH, 1947.

201 Koller, Christophe, *"De la lime à la machine". L'industrialisation et l'Etat au pays de l'horlogerie. Contribution à l'histoire économique et sociale d'une région suisse*, Courrendlin : CSE, 2003, p. 378.

202 Koller, Christophe, *"De la lime à la machine". L'industrialisation et l'Etat au pays de l'horlogerie. Contribution à l'histoire économique et sociale d'une région suisse*, Courrendlin : CSE, 2003, p. 379.

203 *Société générale de l'horlogerie suisse SA. ASUAG. Historique publié à l'occasion de son vingt-cinquième anniversaire, 1931–1956*, Bienne : Arts graphiques SA, 1956, p. 29.

204 AFS, E 7004, 1967/4.22, *L'industrie suisse de la boîte or*, Cartelor, October 1931.

205 *Feuille fédérale*, 1931, p. 312.

206 *Société générale de l'horlogerie suisse SA. ASUAG. Historique publié à l'occasion de son vingt-cinquième anniversaire, 1931–1956*, Bienne, Arts graphiques SA, 1956.

207 *Feuille fédérale*, 1931, p. 213.

208 *Société générale de l'horlogerie suisse SA. ASUAG. Historique publié à l'occasion de son vingt-cinquième anniversaire, 1931–1956*, Bienne : Arts graphiques SA, 1956, p. 59.

209 *Société générale de l'horlogerie suisse SA. ASUAG. Historique publié à l'occasion de son vingt-cinquième anniversaire, 1931–1956*, Bienne : Arts graphiques SA, 1956, pp. 61–62.

210 Pellaton, Jean, *Centenaire de la fabrication de l'assortiment à ancre au Locle, 1850–1950*, Le Locle, FAR, 1950.

211 Nicolet, Georges, *Au cours du temps. Nivarox-FAR, 150 ans d'histoire*, Le Locle, Nivarox, 2000.

212 Linder, Patrick, *Au coeur d'une vocation industrielle : les mouvements de montre de la maison Longines : (1832–2007) : tradition, savoir-faire, innovation,* Saint-Imier : Edition des Longines, 2007.

213 *Société générale de l'horlogerie suisse SA. ASUAG. Historique publié à l'occasion de son vingt-cinquième anniversaire, 1931–1956*, Bienne : Arts graphiques SA, 1956, p. 77.

214 *Feuille fédérale*, 1950, p. 76.

215 The total production of chocolate in Switzerland, which experienced a high growth since the end of the 19th century, increasing from 3140 tons in 1900 to 16253 tons in 1920, was constantly decreasing in the interwar period due to direct investments abroad (6339 tons in 1930 and 327 in 1940). It recovered the 1920 level only in the 1980s. Pfiffner, Alfred, "Chocolat", *DHS*, <www.dhs.ch> (site accessed the 25 June 2009).

216 Klauser, Eric-André, "Sydney de Coulon", *DHS*, <www.dhs.ch> (site accessed the 24 June 2009).

217 *Feuille fédérale*, 30 September 1936, p. 702.

218 Joseph, Roger, "La naissance de la paix du travail", in *L'homme et le temps en Suisse, 1291–1991*, La Chaux-de-Fonds : IHT, 1991, pp. 259–264.

219 *Feuille fédérale,* 1931, p. 204.

220 Art. 1 of the convention, quoted by Joseph, Roger, " La naissance de la paix du travail", in *L'homme et le temps en Suisse, 1291–1991*, La Chaux-de-Fonds : IHT, 1991.

221 *Feuille fédérale*, 1960, p. 1491.

222 *Feuille fédérale*, 1951, pp. 116–117. Unknown data for the years 1951–1959.

223 *Feuille fédérale*, 1950, p. 106.

224 Pasquier, Hélène, *La "Recherche et Développement" en horlogerie. Acteurs, stratégies et choix technologiques dans l'Arc jurassien suisse (1900–1970)*, University of Neuchâtel, PhD thesis, 2007, p. 36.

225 Pasquier, Hélène, *La "Recherche et Développement" en horlogerie. Acteurs, stratégies et choix technologiques dans l'Arc jurassien suisse (1900–1970)*, University of Neuchâtel, PhD thesis, 2007, p. 45.

226 Quoted by Pasquier, Hélène, *La "Recherche et Développement" en horlogerie. Acteurs, stratégies et choix technologiques dans l'Arc jurassien suisse (1900–1970)*, University of Neuchâtel, PhD thesis, 2007, p. 45.

227 Landes, David S., *Revolution in time, clocks and the making of the modern world*, Cambridge: Harvard University Press, 1983, p. 385.

228 Uttinger, Hans W. and Papera, D. Robert, "Threats on the Swiss Watch Cartel", *Western Economic Journal*, 1965, p. 209.

229 Seiko Institue of Horology, national production statistics.

230 Rieben, Henri, Urech, Madeleine and Iffland, Charles, *L'horlogerie et l'Europe*, Neuchâtel: La Bacconière, 1959, p. 53.

231 Koller, Christophe, *"De la lime à la machine". L'industrialisation et l'Etat au pays de l'horlogerie. Contribution à l'histoire économique et sociale d'une région suisse*, Courrendlin : CSE, 2003, p. 337.

232 Pasquier, Hélène, *La "Recherche et Développement" en horlogerie. Acteurs, stratégies et choix technologiques dans l'Arc jurassien suisse (1900–1970)*, University of Neuchâtel, PhD thesis, 2007, pp. 172–179.

233 SSA, SMUV, 04-0063, statistics on roskopf watches' production.

234 Trueb, Lucien F., *125 ans de chronométrage Longines : l'équité dans la mesure du temps, l'élégance dans le sport*, St-Imier : Edition des Longines, 2003, p. 15.

235 Marion, Gilbert, "Alfred Lugrin", *DHS*.

236 Pasquier, Hélène, *La "Recherche et Développement" en horlogerie. Acteurs, stratégies et choix technologiques dans l'Arc jurassien suisse (1900–1970)*, University of Neuchâtel, PhD thesis, 2007, p. 73.

CHAPTER 4
Liberalization and globalization (1960–2010)

The 1960s appear as a new rupture in the history of the Swiss watch industry, which faced a triple mutation, commercial, organizational and technological. At first, this decade was marked by the establishment of new competitors on the world market (Japan, USA, USSR, France and Germany), whose success was essentially based on the mass production of standardized products, and called into question the dominant position of Switzerland. The necessity to reinforce the competitiveness of Swiss watch enterprises led to the end of the cartel; the watch industry reorganized its structures, characterized by the liberalization of cartel's constraints and the international expansion of manufacturing. Finally, the last mutation was technological. During this decade the quartz watch was developed, a technical revolution which significantly transformed the watch business, as from then on the issue was no longer the ability to manufacture watches, but rather the ability to sell them.

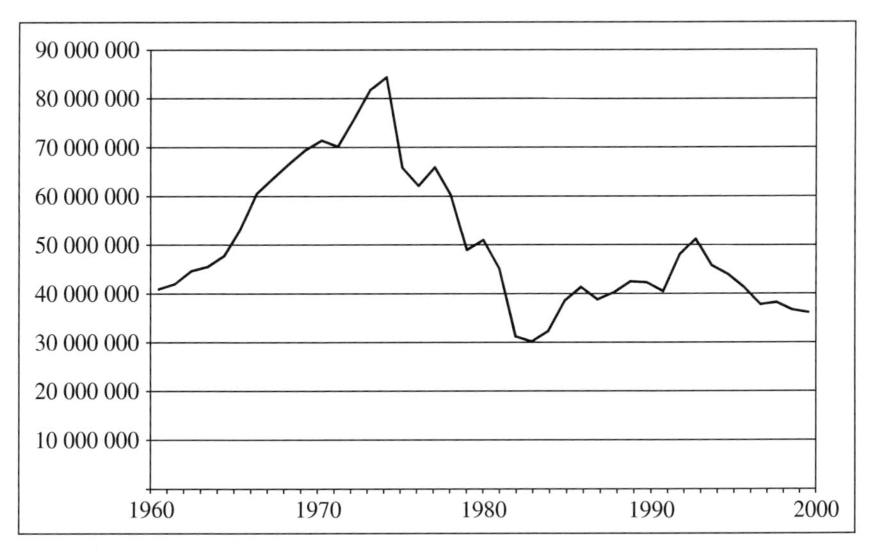

Figure 12: Swiss watches exports, volume (1,000 pieces), 1960–2000

Source: *Statistique du commerce de la Suisse avec l'étranger*, Berne : Département fédéral des Douanes, 1960–2000.

This triple mutation led to a deep transformation of the Swiss watch industry, marked by ten years of crisis and stagnation. The evolution of watch exports between 1960 and 2000 is an excellent illustration of this shift. These four decades can be divided into three periods. First, there was a high growth phase based on an expanding world demand (1960–1974). Freed from the restrictions of the cartel, Swiss watch enterprises developed their means of production and exported huge numbers of cheap watches to the whole planet. In fourteen years, the volume of exports doubled – from 41 million to 84 million pieces – while its value in francs nearly multiplied by three – from 1.3 to 3.7 billion francs.

This high growth of exports relied on the production of different kinds of watches. At first, one must emphasize the steady increase of cheap mechanical watches, the so-called roskopf or pin-lever watches. These products rose from 31.8% of the total volume of exported watches in 1945 to 44.4% in 1960 and peaked at 49.5% in 1972. They were produced by many companies, among which BFG Baumgartner Frères SA, founded at Grenchen in 1916, was one of the biggest. However, due to their low quality, roskopf watches had a very low weight within the overall value of Swiss watch exports – 15.5% in 1960 and 16.5% in 1972.[237] Secondly, there were high quality mechanical watches. In this period, the main object of competition between firms became self-winding watches, which embodied high precision and perfection until the quartz revolution.[238] Rolex, which patented its self-winding watch in 1931, was followed by Omega (1943), Longines (1945) and then Eterna (1948). In the following decades, many other firms launched their own models, sometimes accompanied by an incremental innovation, such as the self-winding alarm watch of Jaeger-LeCoultre (1956) or the self-winding chronograph of Heuer (1969). Even the Japanese challengers developed their own self-winding watches as early as the 1950s. Seiko marketed its first model in 1959 with the objective of competing with Swiss watchmakers on the world market.[239]

Indeed, despite the huge growth of its watch exports after 1945, Switzerland was facing new competitors and did not see its share of the world market increase. Rather, in the 1960s it was in stagnation at around 40–45% of the total market (see Table 18). The post-war boom favored the development of rival nations, such as Japan (13.5% of the world market's share in 1970), the USSR (12.4%) and the United States (11.0%). This euphoria period peaked in 1974 and was suddenly interrupted by the world crisis, which was the opportunity for Japanese and Hong Kong watch makers to establish themselves on the world market, with their precise and cheap quartz watches. The Swiss watch industry then entered a decade of crisis (1975–1984). Its exports were dropping and, at 30 million pieces in 1983, they reached their lowest number since 1950. Roskopf watch makers were the first to experience the crisis due to the lack of competitiveness of these low quality goods against quartz watches. Between 1974 and 1985, the export of these watches dropped from 37.6 million of pieces (47.0% of the total volume) to 1.2 million (3.3% of the total volume).[240] Baumgartner Frères SA closed down in 1982.[241] The crisis was however not limited to roskopf watch producers but spread to all the industry. It actually accelerated the process of restructuring the industrial district towards concentration, that had been occurring since the beginning of the 1960s and whose features were the loss of autonomy of numerous family firms and the birth of group enterprises. Production was reorganized within these new companies, mainly the Société Suisse de Microélectronique et d'Horlogerie (SMH, the Swatch Group from 1998), which was founded in 1983, and oriented from then on towards the top of the range and luxury watches which are the main features of the industry nowadays.

4.1 Decartelization

The establishment of new watchmaking nations on the world market challenged the dominant position of Switzerland. The high competitiveness which appeared in the 1950s led the managers of some Swiss watch firms to request the suppression of the cartel and the liberalization of the industry. The policy to maintain an industrial district made up of hundreds of independent small and medium sized firms was severely criticized. Indeed, the increased competitiveness of the world market made it necessary to adopt a new production system (standardization of products, mass production and assembly-line work) largely unknown until then in the Swiss watch industry, as well as the international expansion of manufacturing, both of which could not have been adopted within the decentralized and atomized structure of the industry. In the message it addressed in 1970 to the Federal Assembly, in which it proposed the complete liberalization of the system, the Federal Council declared that *"this structure hinders in particular the efforts to adapt to new production methods. These methods, which include the entirely automatic production of parts and the assembly of movements by assembly-line work, involve more and more considerable investments and, as a consequence, a larger and larger mass production."*[242] The concentration of enterprises and the delocalization of some low value production activities necessitated the liberalization of the watch industry. Moreover, it must be emphasized that the political and social role affected by the cartel in 1930s was no longer relevant. In a talk given in 1959 to the representatives of the watch organizations, the general secretary of the Federal Department of Public Economy, Karl Huber, emphasized the point of view of the authorities: *"The Department recognizes perfectly the importance these problems have for the maintenance of an independent middle class and of the traditionally horizontal structure of our watch industry. But this objective does not have to be considered as a goal in itself, it must be examined in relation to the whole. Eventually, the determinant is to maintain the competitive capacity of all of the Swiss watch industry, which is an indispensable condition to reach all the other objectives which have been proposed."*[243]

In order to promote the modernization and the liberalization of the Swiss watch industry, the FH appointed Gérard F. Bauer (1907–2000) as its new president in 1958.[244] Former diplomat at Paris since 1945, assigned to the Organization for Economic Cooperation and Development (OECD) and to the European Coal and Steel Community (ECSC), he run the FH until 1977 and played a key role in the transformation of the watch industry. Relinquishing the cartel was at the heart of watch policy at the end of the 1950s. However, this did not have unanimous support within watch industry circles, unable to present a common position to the federal authorities at the end of the decade.[245]

The most reluctant were the producers of parts and movements, grouped together within the UBAH and the company Ébauches SA. The ASUAG was particularly virulent in its open criticism of the proposals to give up the production licenses and the control of the export of ébauches. According to it, the key issue was to maintain its monopolistic position as an ébauches supplier in the Swiss watch industry. As for the other subcontractors, their attachment to the cartel can be explained by the lack of international competitiveness of these sectors. In particular, the watch case and dial makers defended the principle of compulsory prices and the banning of supply of parts mass produced abroad. Their main representative at the Federal Assembly, the Radical (right) national councilor Simon Kohler, intervened at many times in the parliamentary debate on the reform of the watch cartel, in December 1960. Denouncing the *"Manchester-type liberalism one would like to reestablish"*, he declared that *"it may be plausible that some captains of industry, such as Mr. Schmidheiny,*[246] *our colleague, inspired by their own success, would not be inclined, even at least to show comprehension for the watch industry, whose particularities are unique and which is so often subjected to the smallest economical fluctuations."*[247] Contrary to the complete watch makers, the parts suppliers were positioned on a national market and feared that liberalization would, above all, be of benefit to foreign competitors.

The main promoters of decartelization were indeed complete watch makers. Present on the world market, they were facing the foreign challengers and claimed the liberalization of their industry was needed in order to allow the modernization of their means of production. In the end, liberalization won over a growing number of supporters, in the overall context of a trend towards the decartelization of the Swiss economy in the 1960s.[248] A first revision of the legislation on the watch cartel was adopted by the Federal Assembly in 1961. Notwithstanding, liberalization went smoothly, with a transitional system between 1962 and 1970, during which controls on exports of movement parts and movement-blanks remained in place. Likewise, up until 1965, the State maintained formal production controls on company start-ups and recruitment, even though such measures were applied very loosely. Finally, the watch industry was fully liberalized in 1971 and the Swiss government divested itself of its stake in ASUAG in 1984.[249]

Maintaining control over Swiss production

Yet the end of the watchmaking cartel did not mean complete liberalization of the watchmaking sector in Switzerland. Two types of measures were adopted along with decartelization: technical controls for watches in 1962 and the protection of the *Swiss Made* label in 1971. These two steps were designed to ensure a certain quality of Swiss production, in order to maintain the comparative advantage generated by the reputation of Swiss watches on a liberalized, competitive market. Introduced in 1962, technical controls for watches reflected a widespread fear in watchmaking circles of an increase in the number of watch manufacturers who would churn out lower-quality products in order to benefit from prevailing low market prices. Such a policy, over time, could well be detrimental to the watch industry as a whole. It was therefore necessary to introduce technical controls for watches and watch movements manufactured in Switzerland, *"with a view to preventing the export of watchmaking products that could be gravely prejudicial to the renown of the Swiss watch industry".* [250] In practice, however, only a fraction of output was actually subjected to quality controls. From 1972 to 1979, the Institut pour le contrôle officiel de la qualité dans l'industrie horlogère suisse (Institute for offi cial quality control in the Swiss watch industry), an independent body under State control, verified a paltry 1.8 million watches in all, that is, a mere 0.3% of all watch exports.[251] These technical controls were combined with the *Swiss Made* label for watches, adopted by legislators at the time of liberalization in 1971. The label stipulated a number of criteria, the main requirement being that at least half of the value of the components of the movement must be produced and assembled on Swiss soil. Consequently, half of movement's parts and of all the external parts (case, dial, hands, strap, etc.) can be supplied from foreign plants. This is a very pragmatic measure with a twofold objective: keeping employment in Switzerland and strengthening the competitiveness of Swiss watch companies by allowing relocation abroad – mainly in Asia – of low value-added production activities.

4.2 The quartz revolution

The 1960s and the 1970s were marked by a technological revolution in the watch industry: the creation of quartz watches. The development of a marketable electronic watch was a key issue from both technical and economical points of view during this period. The Swiss, American and Japanese watch enterprises launched a strategy of research and development (R&D) whose aim was to enable them to innovate in this field.[252] The first prototypes of quartz watches were completed in 1967 by the Centre Électronique Horloger (CEH), in Switzerland, and the company Seiko, in Japan, which was the first to commercialize this innovation in 1969. Then, prototypes were presented successively by Longines (1969), Hamilton (1970), Omega (1970) and Radio Corporation of America (RCA, 1971). However, for Swiss companies, the largest problems occurred with the mass production of these watches, not in developing prototypes.

In Switzerland, the atomized structure of the industry largely conditioned the way R&D was pursued. Three approaches were adopted to acquire this new technology: the creation of a community research center (Centre Électronique Horloger, CEH), collaboration with American and European electronic firms, and R&D within single enterprises.[253]

The CEH was founded in 1962 at the instigation of the FH.[254] Its main shareholders were both private firms (Ébauches SA, ASUAG, Longines, Omega, etc.) and the principal watch organizations (the Fédération Horlogère and the Chambre Suisse d'Horlogerie). The objective of this center was the development of an electronic watch which would allow the Swiss watch industry to compete with the American company Bulova, which released the first prototypes of tuning fork watches in 1955 (the Accutron). However, the CEH did not have the aim of marketing this innovation, but rather to supply it to private enterprises. Like Longines and Omega, these firms were largely doing in-house R&D and had an ambiguous attitude towards the CEH. Their participation in community R&D was a part of a technology scouting strategy.[255] The work of the CEH made it possible to it to present a prototype of a quartz watch, the Beta 21, in 1967. The production of a first series of 6,000 movements was taken in charge by three firms, CEH, Ebauches SA and Omega, which offered electronic watch movements to their customers from 1972.

The collaboration with American and European electronic firms occurred on the community level as well as within private companies. At first, on the community level, one should mention the creation in 1966 of the joint venture Faselec, a stock company created by two watch companies (FH Electronic Holding, Ebauches SA), three Swiss electric companies (Brown Boveri Co., Landis & Gyr, Autophon) and the Dutch multinational Philips.[256] The aim of this firm was to develop R&D in the field of semi-conductors and chips, in order to reduce the dependency of the Swiss watch industry on American electronic companies. A first prototype of a digital quartz watch was presented to the public at the 1973 Basel Fair. As for the CEH, it proceeded to a major change in 1972, with the arrival of the company Bulova in the capital. From then on the American watch company was integrated into the community Swiss R&D.

Finally, R&D occurred also within some companies. The collaboration with foreign partners was often essential. Thus, Ébauches SA, already involved in CEH and Faselec, pursued its own researches. From 1968 it collaborated with Bulova, with which it signed an agreement on the production under license and the commercialization in Switzerland of tuning fork watches. It also collaborated with Texas Instruments and Brown Boveri, which made it possible to acquire know-how respectively in chips and liquid crystal technologies. As for the watch manufactures, they were doing their own R&D and marketed several quartz watches at the beginning of the 1970s.

Thus, the Swiss watch industry mastered quartz technology as soon as the end of the 1960s, but it was not able to launch products competitive with Japanese, American and later Chinese watches on the market. The inability to transform a technical innovation into a marketable product can largely be explained by the fragmented structure of the industry in Switzerland. R&D appeared to have been extremely dispersed and unlinked to marketing projects, unlike the main rival of Swiss watchmakers, the Japanese company Seiko, for which the realization of a marketable quartz watch was a strategic choice of the firm as early as the beginning of the 1960s.

The prototypes created by the CEH did not fulfill any demand from complete watch makers who would have desired to launch such new watches. They were, on the contrary, the result of a will, largely shared within elite watchmakers, to master new technologies rather than a real desire to sell new kinds of watches. Besides, the principal companies which had R&D activities in electronic watches (Longines, Omega and Girard-Perregaux) were manufactures having an old tradition and ancient history with most of the production oriented towards top-of-the-range and luxury. Even if it appeared as an essential innovation, notably in the field of sport's timing, which was itself a key issue in communication, quartz technology was not thought by these companies as a revolutionary product. Moreover, they lacked the production facilities necessary to make an effective shift to the large-scale manufacture of electronic watches.

Nevertheless, the development of quartz watches was a commercial and technological revolution. Until the 1960s, the issue for watchmakers was to have access to the movement of the watch. It was a very complex mechanism whose production was controlled by the ASUAG and its various sub-holding companies. In attempting to put an end to *chablonnage* in the interwar period, the watchmaking milieu had adopted a cartelization policy whose objective was precisely to control the world watch market through the control of the production and the sale of movements. Yet, the appearance of the electronic watch at the end of the 1960s put an end to the supremacy of ASUAG. The invention of quartz watches liberalized and democratized the access to the motor of the watch: from then on, any entrepreneur, anywhere in the world, could acquire electronic watch movements for a very cheap price and engage in the business. That was exactly what caused the success of Chinese makers from Hong Kong. They established themselves in the low end range of the market during the 1970s and the 1980s, and became the third largest watch producers after Switzerland and Japan. The value of watch exports from Hong Kong rose from 208 million HK$ in 1970 to 6.6 billion in 1980 and peaked at 19.1 billion in1990.[257] In terms of market share, the advent of the quartz watch called into question the dominant position the Swiss had held over the world market since the second part of the 19th century. A huge expansion of the world market occurred, from 177 million watches in 1970 to 531 million in 1983, an evolution which mainly benefited the Asian newcomers. In 1983, the largest watch producers were Hong Kong (38%) and Japan (23%), in front of Switzerland (9%), which was relegated to third.[258]

4.3 The origins of the "watchmaking crisis"

The period of crisis and reorganization, which the Swiss watch industry was coping with between the middle of the 1970s and the middle of the 1980s, and which was commonly called the "watchmaking crisis", is usually considered as a direct consequence of the quartz revolution and the inability of Swiss industrialists to adopt this innovation. Yet, even if it is true Swiss enterprises faced huge difficulties in the shift towards the industrialization of electronic movements, as Hélène Pasquier has shown,[259] Japanese competitors took several years to impose their technology. Omega and Longines mastered quartz technology perfectly and as early as their Japanese rivals. Even if Seiko was the first company in the world to launch a quartz watch, on the Christmas Day 1969, it was above all a marketing operation planned by the management.[260] In the following years, Seikos quartz watch production was still very low (3,000 pieces in 1971 and 64,000 in 1972, that is less than 0.5% of the overall volume of production) and it was only in 1979 that it exceeded that of mechanic watches.[261]

The crisis experienced by the Swiss watch industry was thus only the indirect result of a technological change linked to the nature of the product. It appears to be more the consequence of structures unsuitable to globalized capitalism. While maintaining an industrial base composed of hundreds of small and medium sized firms, interdependent but autonomous,

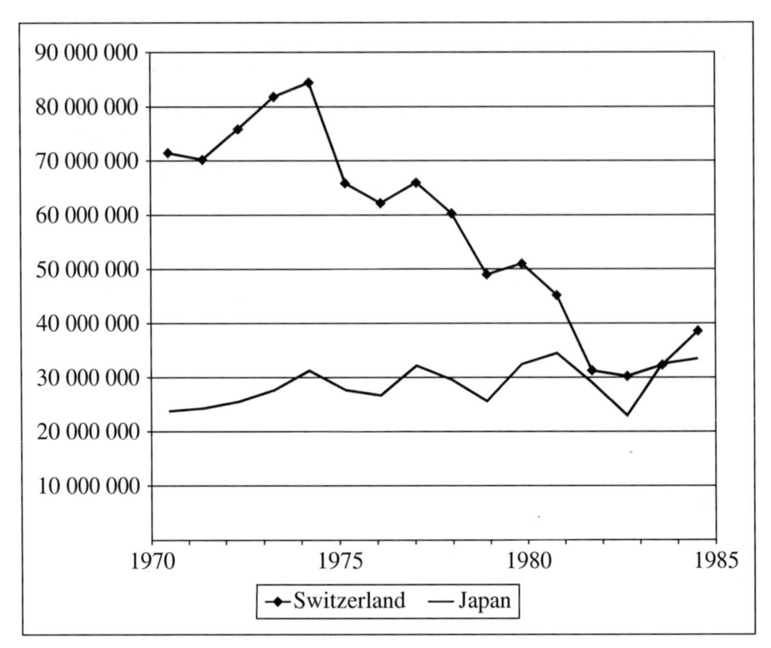

Figure 13: Production of Swiss and Japanese mechanical watches, volume, 1970–1985

Sources: *Statistique du commerce de la Suisse avec l'étranger*, Berne : Département fédéral des Douanes, 1970–1985 and statistics communicated by the Seiko Institute of Horology, Tokyo.

Note: For Switzerland, the volume of electronic watches being negligible before the Swatch, the volume of export watches was taken into account. For Japan, the estimations of the years 1972–1974 are based on an extension at the national level of the production of electronic watches by Seiko.

the watchmaking cartel (1934–1971) delayed the industrial concentration which was necessary for a rationalization of production and marketing on a competitive world market. The strength of Seiko, in the 1960s and the 1970s, was precisely the ability to mass-produce and mass-distribute Swiss-like quality watches.[262]

The evolution of the production of mechanical watches by both countries during the period 1970–1985, the advent of quartz watches, sheds light on this phenomenon. While Switzerland, after having reached a peak of 84.4 million pieces in 1974, entered an inexorable fall until the middle of the 1980s, marked by an annual average production of 31.3 million pieces for the years 1982–1984, Japan did not experience any crisis in the production of mechanic watches. Of course, there was a change from

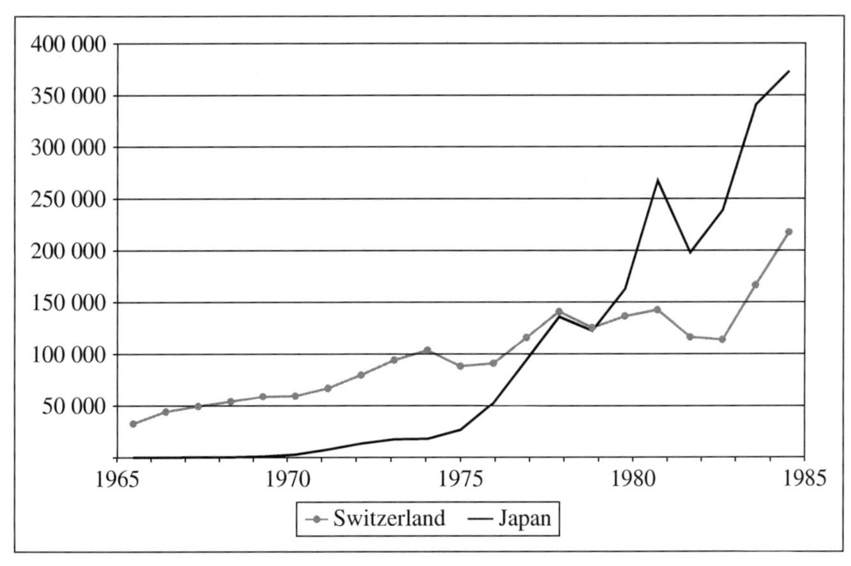

Figure 14: Import of complete watches in the United States, in thousands of dollars, 1965–1980

Source: U.S. Department of Commerce, Bureau of the Census, *U.S. Imports commodity by country*, 1965–1980.

the previous decades characterized by a very high expansion (0.7 million pieces in 1950 and 7.1 million in 1960).[263] Nevertheless, although it was the pioneering nation in the development, and especially the marketing of quartz watches, Japan continued to produce mechanical watches and even had some growth, although small, in this sector during the 1970s: it amounted to 23.8 million pieces in 1970, 27.7 million in 1975 and 32.4 million in 1980. In 1984, the production of mechanical watches in Japan (32.5 million) exceeded Swiss output (32.3 million) for the first time. Thus the appearance of the quartz watch is not a sufficient factor to explain the "watchmaking crisis".

The main problem for the Swiss watch industry in the 1970s was its lack of competitiveness on the world market. Of course, it had an appreciable advantage in terms of image and sales volume, but its products were not competitive. Swiss watches were too expensive, especially due to a lack of rationalization of production and of a marketing strategy organized and coordinated at the global scale. The American market, the most important in the world, is a very good illustration (see Figure 14).

In 1965, Switzerland had a quasi-monopoly with 92.2% of complete watch imports, while Japan's contribution only amounted to 0.2%. In absolute numbers, the import of Swiss watches shows a continuous growth until 1985, despite the fall in 1975–1976. However, their relative share was constantly decreasing: 83.1% in 1970, 58.8% in 1975 and only 22.2% in 1980. In fifteen years Swiss watches dropped from a monopoly situation to less than a quarter of the market. The main reason for this shift was the growth of Japanese watch imports, which became exponential during the 1970s, and whose value exceeded that of Swiss watches in 1980. In relative numbers, this growth was already appreciable at the end of the 1960s. Indeed, the share held by Japanese watches amounted to 4.0% in 1970, 17.6% in 1975 and 26.5% in 1980. Yet, the success of Japanese watches on the American market was not only a consequence of a progressive technological revolution. It was a result of both the set up of a mass-production system for the manufacture of good quality mechanical watches in Japan and of a very well planned expansion into the American market (distribution, advertising and after sales service). For many small Swiss watchmakers, not used to competitive markets, the arrival of Seiko in America was a real shock, reinforced by the fact that Swiss watch exports were oriented towards low quality watches (roskopf and pin lever watches) at the end of the 1950s. Rejected by American importers and distributors, who preferred to deal with Japanese watchmakers, they lost a substantial part of their gross sales and were unable to react.

The situation observed in the United States was similar on all the markets. In Hong Kong, the nerve center for trade in the Far East, the same phenomenon can be observed, even if precociously because the Japanese had already challenged Swiss domination during the 1960s. Indeed, the Swiss share of watch imports in the British colony (by value) decreased from 89.1% in 1960 to 75.8% in 1965 and to 54.4% in 1970. At the same time, the Japanese share rose from 0.7% to 13.4% and 33.3%.[264] In the secondary markets, such as Latin America and the Middle East, Japanese watchmakers also arrived in the second part of the 1960s and called into question the quasi monopoly exerted until then by the Swiss companies.

Finally, there was a last element which historians of the watch industry hardly tackled in their explanation of the watch crisis and which was nonetheless a key element: the monetary factor. A comparative analysis of the exchange rate trends for the Swiss franc and Japanese yen against the

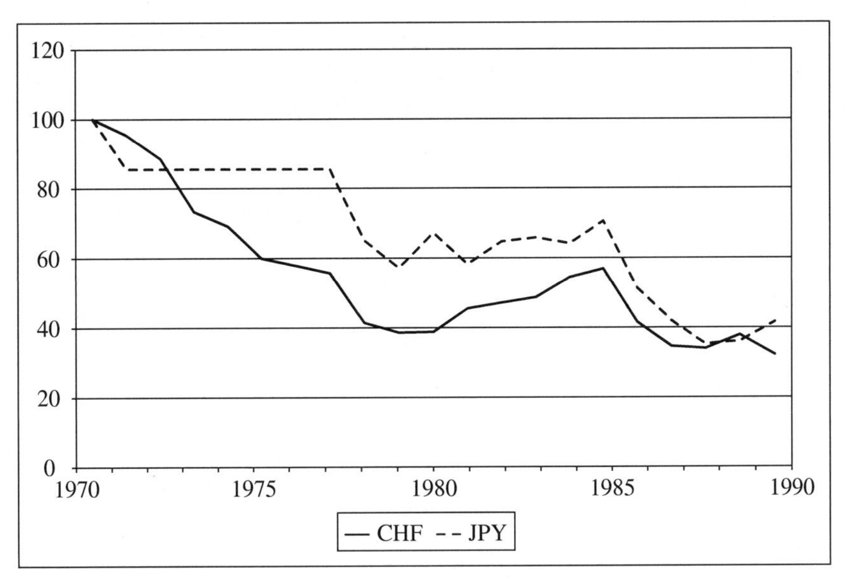

Figure 15: Exchange rate trends for the Swiss franc (CHF) and the Japanese yen (JPY) against the US dollar (100 = 1970), 1970–1990

Source: *For CHF,* Official statistics of the Swiss National Bank, <www.snb.ch> (site accessed 6 July 2009); *for JPY, Nihon chouki toukei souran, Tokyo: Nihon tokei kyokai, 1988, 18–8.*

US dollar in the 1970s and 1980s helps us understand better the unfavorable impact of the monetary factor for Swiss watchmakers (Figure 15).

The international monetary system experienced a deep shift in 1973 with the end of the fixed currency system and the introduction of a floating exchange rate system. As far as competitiveness was concerned, this change had important consequences for the watch business because it caused the Swiss franc to appreciate sharply, raising the value of Swiss watches on the world market, particularly in the US, precisely at a time when Japanese competitors were growing stronger. Between 1970 and 1979, the dollar lost more than 60 per cent of its value against the Swiss franc. At the same time, however, Japan continued to benefit from a fixed exchange rate with the United States until 1977, although the yen was revalued from 360 (its fixed parity since 1949) to 308 yen for 1 dollar in 1971. Obviously, this stability is the major reason why Japanese watchmakers did not suffer from the oil shock: indeed, American imports of Japanese watches grew from 17.9 million dollars in 1974 to 26.5 million in 1975 and 52.9 million in 1976.[265] For the Swiss watchmakers, the high cost of the Swiss franc had disastrous effects: it strengthened the lack of competitiveness of

Swiss watches on the world market. It was particularly true in the United States where in 1975, for the first time since 1960, the value of imported Swiss watches dropped to about 15 million dollars, while the total value of American watch imports continued to increase – from 140 million in 1974 to 150 million in 1975.

Afterwards, even though the dollar lost ground against the yen since 1977, the drop was less brutal than for the Swiss franc. Until 1985, the difference between both currencies towards the US dollar was extremely favorable for Japan, thereby sustaining this country's high growth on the American market. Finally, following the Plaza Agreements (1985), the yen firmed up against the dollar and Japan lost a major competitive advantage against Switzerland, especially since 1987, precisely when the restructured Swiss watch industry undertook its reconquest of the world market, with the Swatch benefiting from an exchange rate that had become favorable in relation to the yen.[266]

The lack of competitiveness of the Swiss watch industry on the world market during the 1960s led to some attempts to reform the industrial structures and to progressively abandon the cartel, in order to enable a rationalization of production and distribution. The concentration into groups of enterprises and the globalization of production were the main features.

4.4 Industrial concentration and the appearance of watch groups

Until the second part of the 1960s, industrial concentration in the Swiss watch industry was very limited due to the cartel's policy. The two biggest groups were ASUAG, created with the aim of controlling the production of watch movements, and SSIH (Omega and Tissot), founded in 1930. The end of the cartel made possible a major mutation of structures, characterized by a large wave of mergers and the birth of watch groups. This redevelopment can be explained by changing causes at the time, but there was a continuous trend which is still going on at the beginning of the 21st century. First, in the second part of the 1960s and the 1970s, the concentration of factories and firms had above all the objective of setting up mass production systems. The economies of scale resulting from concentration enabled Swiss enterprises to improve their competitiveness against the Americans and Japanese on the

Table 19: Number of enterprises and employees in the Swiss watch industry, 1950–2000

	1950	1960	1970	1980	1990	2000
Enterprises	1863	2167	1618	861	572	575
Employees	60,239	74,216	89,448	46,998	33,923	37,334
Employees/enterprise	32.3	34.2	55.3	54.6	59.3	64.9

Source: Convention patronale, *Recensement 2007*, La Chaux-de-Fonds : CPIH, 2008, p. 13.

world market. Second, the mergers of the 1980s and 1990s were essentially the result of a new marketing strategy born from the crisis. As the Swatch Group embodied it, the key issue was to gather into a group enterprises whose brands were positioned in different segments of the market. The main objective of the concentration was not only economies of scale, but also control of the market. Finally, there is a new trend since the 1990s of a vertical integration of subcontractors within these groups. By doing so, they aimed at ensuring the supply of parts for their own needs.

Thus, while growth of the watch industry was based on the creation of new firms in the 1950s, the number of enterprises involved in this business constantly declined from the 1960s: between 1960 and 1990, about three quarters of them disappeared, and not only during the crisis years. At the same time, a growing industrial concentration can be observed, particularly in the 1960s, with the average number of employees in enterprises going from 34.2 persons in 1960 to 55.3 in 1970 (see Table 19).

The first wave of mergers

The industrial concentration trend was especially strong in the years 1966–1971, a period during which three big groups established themselves as the main watch groups of the country. In 1971 they represented more than a quarter of the watch production of Switzerland (see Table 20). This first wave of mergers, whose objective was the rationalization of production, strengthened the two biggest companies, SSIH and ASUAG.

The SSIH remained the largest watchmaker in the country with regard to the volume of production. It was, at the beginning of the 1970s, the third biggest watch company in the world, behind the American Timex and the Japanese Seiko.[267] As early as the beginning of the 1960s, SSIH launched into an all-out offensive of acquisitions, successively taking over Rayville SA, Montres Blancpain, specializing in jewelled watches for women (1961) and Langendorf Watch Co (whose brand is Lanco), a producer of

Table 20: Main watch groups in Switzerland, 1971

Name	Foundation	Main brands	Production (million pieces)	As a % of the national production
SSIH	1930	Omega, Tissot, Lanco, Aetos, Agon, Buhler, Ferex, Continental	10.2	14.6
Société des Garde-Temps (SGT)	1968	Waltham, Avia, Helvetia, Silvana, Solvil, Titus, Invicta, Sandoz	3.3	4.7
General Watch Co	1971	Certina, Edox, Eterna, Mido, Oris, Rado, Technos, Longines, Rotary	5.3	7.6

Source: "Les concentrations dans l'industrie horlogère suisse", *Annales biennoises*, 1971, p. 50.

lever watches (1965). In 1969, it bought into Aetos, which was a group of companies making cheap watches, and then, in 1971, it became the major shareholder of Economic Swiss Time Holding, a company grouping bottom-of-the-range pin-lever watches, with the objective of competing with American Timex. These numerous acquisitions answered both a will to develop production capacity and a new marketing strategy based on a presence in all segments of the market. In the 1960s, the SSIH adopted a policy of segmentation of its brands. Thus, the two main enterprises of the group, Tissot and Omega, showed distinct profiles in 1970: the volume of their production was quite similar (1.2 million of pieces for Tissot and 1.6 million for Omega) but the value of their gross sales was very different (22 million francs for Tissot and 62 million for Omega).[268]

The second group was General Watch Co. (GWC), founded in 1971 by ASUAG which held 60% of the capital. The decartelization of the Swiss watch industry and the emergence of electronic watches challenged the monopolistic position that ASUAG had enjoyed since the interwar years. It then diversified its activities by taking over watchmakers. The GWC group had a total of 11 companies at the beginning of the 1970s, of which the main ones were Longines, Rotary and Rado.

The third group was the Société des Garde-Temps (SGT), created in 1968 and gathering together various cheap mechanical and quartz watch makers. The SGT was also characterized by an international dimension, as it bought up the Waltham Watch brand and signed an agreement with the Elgin Watch Co. for production under license in 1973. It was the biggest direct investment made by a Swiss watchmaking company abroad. However, even if these names were undoubtedly prestigious because of their long history, both these enterprises, oriented to the low end of the range and the domestic American market, were no longer the industrial giants that they used to be. Actually, the SGT bought access to the American market and to a system of distribution through these companies.[269] A rationalization of production was realized in 1971 with the centralization of movement manufacture. However, the exclusive orientation towards cheap watches created severe problems of competitiveness and finally led to the bankruptcy of the group at the beginning of the 1980s.

Finally, one should mention the creation in 1966 of a financial holding company, Chronos Holding SA, founded at Biel with a capital of 21.15 million francs.[270] It possessed its own watch group, Synchron SA, created in 1968 with a capital of 2.4 million francs and including the brands Cyma, Ernest Borel and Doxa, whose volume of production amounted to 250,000 pieces in 1971. At the same time, it had some financial participation in SGT (14% of the capital in 1971) and in the group Saphir SA, which gathered together Favre-Leuba and Jaeger-Lecoultre (23% of the capital). In liquidation in 1978, Synchron SA closed its factories and sold its brands to the family firm Aubry Frères SA which was at the time adopting a strategy of expansion through the takeover of other companies.[271]

The birth of the Swatch Group[272]

The crises the Swiss watch industry was facing in the second part of the 1970s initially reinforced the positions of the two biggest watchmaking enterprises of the country, ASUAG and SSIH: in 1979, they represented about half of the employment in the watch industry in Switzerland.[273] Yet this power must be put into perspective, because it was largely the consequence of the great difficulties other companies were experiencing. In reality, both ASUAG and SSIH had very severe industrial and financial difficulties. And in the end they owed their survival only to the support of the big banks of the country.

ASUAG had difficulties in dealing with its growth and the diversification of its activities. While it had founded its development from the 1930s on a monopolistic position as a producer and distributor of ébauches and parts, the advent of the quartz watch challenged this traditional activity. In 1971, it tried to diversify into the production of complete watches, with the creation of the General Watch Co. (GWC), a strategy which had a strong impact on its finances and management. Between 1970 and 1974, the balance sheet of the ASUAG grew from 54.5 million francs to 234.1 million. Obviously gross sales also grew, increasing from 760 million francs in 1970 to a peak of 1.4 billion francs in 1974. Nevertheless, the growth was very fragile. ASUAG relied more and more on external capital, which had made possible the setting up of GWC. It amounted to 28.3% of the balance sheet in 1974 against 23.7% in 1970. Then, when the Swiss watch industry entered recession, the gross sales of the ASUAG dropped to less than 1.2 billion francs each year during the period 1975–1978 and dependence on capital from banks strengthened.[274]

As for the SSIH, it was also facing an important crisis after 1974 due to its rapid and large diversification strategy, especially in the low end of the market (Economic Swiss Time) which was unable to compete with Japanese and American quartz watches. Between 1974 and 1982, SSIH was collapsing: its sales dropped from 12.4 million to 1.9 million watches, the gross sales from 733 million to 537 million francs and the number of employees from 7,300 to 3,400 persons.[275] The banks which supported the group, giving medium term loans, actively engaged in the restructuration of SSIH. A restructuring committee, led by the Union Bank of Switzerland (UBS), the Swiss Bank Corporation (SBC) and Credit Suisse, was set up in 1980. The next year, a former director of the UBS, Peter Gross, took over the chair of the Board of Directors. Between 1981 and 1983, the banks invested more than 900 million francs in SSIH and ASUAG and were trying to dispose of them out as quickly as possible.[276]

The two main difficulties of both of these groups were the management of inadequately integrated groups and the unsuitability of products for the market. Only a strategy of industrial restructuring and of market rationalization could ensure their survival, and beyond them the survival of all industry in Switzerland. Banks had recourse to the advice of the consultant Nicolas G. Hayek in order to realize this reorganization. Born at Beirut in 1928 and trained at the University of Lyon (France), in 1963

Hayek founded a consulting company at Zurich, Hayek Engineering, active in the counseling of enterprises, and working notably for several watchmakers whose restructuring was a necessity within the framework of the mutations of the Swiss watch industry (liberalization, development of quartz watches and foreign competition).[277] In the report he gave to the banks, Hayek proposed as the main measure the merger of ASUAG and SSIH into a new group, the Société Suisse de Microélectronique et Horlogerie (SMH). It was created in 1983 and gave birth to the largest watchmaking group in the world, which took the name Swatch Group in 1998. Hayek acquired the majority of the capital in 1985 and the following year became chairman and chief executive officer of the Board of Directors. Heading this enterprise, he had the opportunity to carry out an innovative industrial policy which largely contributed to the rebirth of the Swiss watch industry. The principle of this new policy was the primacy of marketing over production: as the quartz revolution made it possible for anyone to manufacture watches, the issue was no longer how to make them, but rather how to sell them. The features of this new industry were the creation of watchmaking groups built on the concentration of old enterprises, which enabled the rationalization of production and distribution networks, as well as the adoption of a marketing strategy controlled at the group level (brands positioning).

This industrial concentration led to a rationalization policy. At first, it concerned production and more particularly watch movements. The R&D activities and the technical departments of the various factories were gradually closed down and centralized within ETA (the new name of Ébauches SA) at Grenchen. Omega stopped producing its own movements in 1985 and Longines in 1988. This rationalization of movement production made possible a better control of costs, thanks to economies of scale, and it allowed the different companies in the group to concentrate on commercial aspects. The definition of a new marketing paradigm was the main feature of the industrial strategy pursued by Hayek. From then on, the different brands of a group were not autonomous enterprises producing specific watches. The result of this rationalization was the launch of very similar products, whose external design (case, dial and hands) and mode of selling were the only variations.

A true product of marketing was also at the origin of the success of Hayek's new watchmaking policy: the Swatch watch. Launched in 1983, this plastic quartz watch was thought of as a cheap mass product, whose particular design was changed according fashion and events. The worldwide success of the Swatch watch enabled the reactivation and development of the entire group. The profits resulting from the Swatch watch were invested in the acquisition of companies to make it possible for the group to be present on all segments of the market. Thus low end products enabled this new policy. In addition to the brands already present in ASUAG and SSIH, the Swatch Group steadily bought up new ones (Breguet, Blancpain, Glashütte and Calvin Klein). The gross sales of the group grew from 2.1 billion francs in 1990 to 2.6 billion in 1995, 4.3 billion in 2000 and 6.4 billion in 2010. It employed 12,800 people in 1990, 19,700 in 2000 and 25,197 in 2010.[278]

The main watch groups in the 2000s

Although the Swatch Group has a largely dominant position in the Swiss watch industry, it is not the only enterprise to have developed an industrial concentration policy based on a marketing strategy. Its two main rivals are the Richemont Group and LVMH. However, unlike the Swatch Group, these are not former watch companies which were reorganized and concentrated within a new company. They are foreign luxury goods groups which acquired Swiss watch companies as part of a diversification strategy. The Richemont Group, which belongs to the tobacco maker family Ruppert (British American Tobacco, BAT), was created in 1988 as a way to diversify the familial business into luxury goods, with the takeover of Baume & Mercier and Piaget in that year, and then of Vacheron Constantin (1996). In 2008, it also owned the brands A. Lange & Söhne, Cartier, Dunhill, IWC, Jaeger-LeCoultre, Montblanc, Panerai, Ralph Lauren, Roger Dubuis, Shanghai Tang, as well as Van Cleef & Arpels. The gross sales of its watchmaking division, not including jewelry watches (Cartier and Van Cleef) and accessories (Dunhill and Montblanc), amounted to 885 million euro in 2005 and to 1.4 billion euro in 2010.[279] As for the French luxury group LVMH, it also undertook a strategy of diversification into watchmaking through the taking over of Swiss companies. In 2008, it notably owned the firms Tag Heuer, Zénith and Hublot. Its gross sales in watchmaking and jewelry amounted to 585 million euro in 2005 and 985 million euro in 2010.[280]

The industrial strategy of the watch groups was not only characterized by the merger of watchmaking companies and the repositioning of brands. One should mention also, since the middle of the 1990s, the strong tendency towards vertical integration. The buying up of subcontractors,

Table 21: Takeover of Swiss case makers by watch groups

Watch group	Case maker	Year
Desco de Schulthess	Queloz SA, Saignelégier	1990
Ebel/Cartier (Richemont)	Cristalor SA, La Chaux-de-Fonds	1992
Tag Heuer (LVMH)	Cortech SA, Cornol	1992
Cartier (Richemont)	Paolini SA, La Chaux-de-Fonds	1993
Audemars-Piguet	Centror SA, Genève	1995
Swatch Group	Favre & Perret, La Chaux-de-Fonds	1999
Fondation Sandoz	Affolter, La Chaux-de-Fonds	2000
Patek Philippe	Calame & Cie, La Chaux-de-Fonds	2001
Mouawad (Robergé)	Mica SA, Les Breuleux	2004
Richemont	Donzé-Baume SA, Les Breuleux	2007

Source: Archives of the Union suisse pour l'habillage de la montre (USH), Biel.

particularly in the *habillage* business (cases, dials and hands), was numerous and had the aim to enable a better control of the supply of parts. For an industry in which image and communication became the key issues in the sale policies, it was a necessity to have control over the design of watches. The example of case makers is representative of this trend: ten of them were taken over by watch groups during the 1990s and 2000s (see Table 21).

The Swatch Group, LVMH and Richemont account for an essential part of Swiss watch industry. Nevertheless, the persistence of a large number of independent firms, supported by a good economic situation, was also a feature from the end of the 1980s. The number of watch companies, which dropped from 861 in 1980 to 572 in 1990'stabilized from then on: there were 575 in 2000 and 593 in 2005.[281] They were mainly small watchmakers which obtained movements from the Swatch Group and who sold their output in few specialized markets. However, among these independent firms, one should mention the existence of some of the most famous companies of the industry, such as Rolex and the luxury watch producers of Geneva.

An independent firm: Rolex

Rolex is a grey area in the history of the Swiss watch industry. While it is obviously, together with Omega, the most famous watch brand in the world, little of its history is known. The manufacturer Rolex belongs to a private foundation which does not publish annual reports, and does not provide any information on the development of the firm, so that it is particularly difficult to understand the path of growth of this phenomenal success story.[282]

The brand Rolex was firstly registered in 1908 at London by Hans Wilsdorf (1882–1960), a trader who was selling watches in the British Empire. In 1915, he moved his business to Geneva for financial reasons and founded the Montres Rolex SA company in 1920 to case-up the movements and market the watches. His main, and soon sole supplier of movements was the Fabrique Aegler, a watch factory founded at Biel in 1878 by Jean Aegler (1850–1891). It produced movements, notably for women's watches, displaying a technical expertise in the manufacture of tiny components that qualified it as a pioneer in the introduction of wristwatches. Furthermore, Aegler soon developed into a factory turning out watches on an industrial scale. It was a modern plant, mechanised since 1898 and with a workforce organised for industrial production – in 1916 it became involved in a violent strike over wage demands by its machine operators, which spread to the other local factories. The Aegler factory remained a focus of heightened social tensions throughout the inter-war period.

Aegler was a family enterprise. One of the founder's sons-in-law, Emile Borer (1898–1967), joined the company as an engineer during World War I. He took charge of developing new techniques, the best known of which is the winding rotor for selfwinding wristwatches in 1931, and became the company's general manager in 1944. The company, which was to become the Manufacture des Montres Rolex, manufactured movements exclusively for Montres Rolex SA in Geneva, which acquired it in 2004.

In the interwar years, Rolex had a very intense period of innovation, including the creation of the Oyster waterproof case, automatic movements for wristwatches and the serial production of chronometers. This technological development enabled Rolex to establish itself after 1945 as the main manufacture of mass-produced chronometer watches.

It is likely the success of Rolex is based on the early and systematic adoption of mass production of quality watches. In other watchmaking firms, low quality watches (pin-lever watches) were usually produced in this way, while manufactures like Omega or Longines were continuing to make a wide range of products. The characteristic of Rolex is its small range of products (essentially the Oyster model in diverse varieties), their high quality (quasi systematic chronometer certificates) and a very limited diversification into quartz watches. The annual statistics of the Contrôle officiel suisse des chronomètres (Official Swiss Chronometer Testing Institute, COSC) brings to light this aspect of Rolex (see Figure 16). During the 1960s, Rolex and Omega presented a similar profile showing a growth of their quality products: their watches amounted at 84.6% of the chronometers certified by the COSC during the decade. However, a sudden shift occurred in the 1970s. Omega nearly disappeared while Rolex pursued its growth, a trend which went on during the 1980s, so that Rolex watches alone amounted at 81.3% of certified chronometers in 1980 and 93.9% in 1990. Its gross sales, not published, were estimated at 200 million francs in 1973 and three billion in 2007.[283] Whereas Swiss watchmaking in general, dealing with a serious crisis, was hesitating between mass-produced cheap mechanical watches, quartz watches and quality goods, Rolex maintained its strategy of mass manufacturing high quality but not too expensive watches. It is itself the example that the lack of competitiveness of the Swiss watch industry on the world market in the 1960s was not a consequence of problems linked to quartz watches but to the incapacity to adopt a systemized mass production system for mechanic watches – and thus to structures inadequate to globalized capitalism.

The exception of Geneva: the evolution of luxury watch makers during the second part of the 20th century

Traditionally focused on jewel-watches and luxury goods, watchmaking in Geneva shows a particular path of development within the Swiss watch industry. Even if it was also exposed to changing circumstances, it stayed at the margin of the general trend towards a more mass-production oriented structure and relied on the persistence of an industrial base made up of small luxury goods' firms. While Rolex integrated top-of-the-range and mass production, the watchmakers of Geneva stayed positioned in the luxury and craft industry segment of the market. It is unknown when the shift towards luxury occurred at Geneva, but the end of the 19th century must have been an important period of change. Facing the American challenge, most of the Swiss watch companies had no choice but to industrialize their production. Few of them chose, on the contrary, to keep manufacturing jewellery watches for wealthy people. These companies positioned themselves to satisfy the needs of a very rich clientele who could afford exclusive status symbols, in exactly the same way that Abraham-Louis Breguet satisfied the elite circa 1800.[284] More recent developments, in particular the individual makers who are now in quite large numbers, have resulted from a dramatic increase in the numbers of wealthy people who can afford such watches and need them to bolster their self-perception. Nevertheless, corporate histories and accessible archives are too rare to provide enough information to be able to fully understand the phenomenon. However, the path of two companies enables to illustrate the main characteristics.

The first example is the company Piaget, whose success was based on the union of a watchmaker and international jewellery big business.[285] The origin of Piaget goes back to a watchmaking workshop set up in 1874 at La Côte-aux-Fées, in the canton of Neuchâtel. It was a small family firm, like dozens of others in the Jura Mountains. It had strong growth in the 1950s and 1960s thanks to the business relationships it established with famous jewelry makers, such as Garrard & Asprey at London, Tiffany at New York and Cartier at Paris, who were all selling

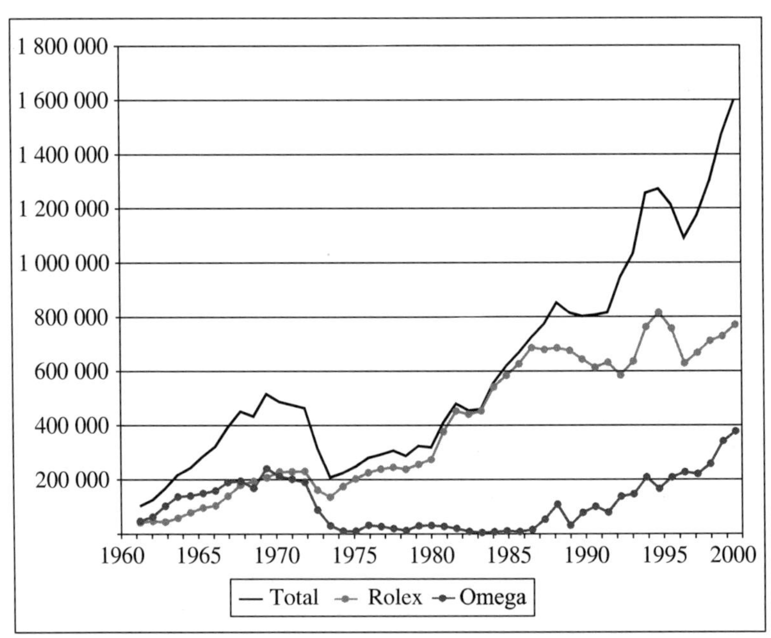

Figure 16: Certificates given by the Official Swiss Chronometer Testing Institute (COSC), 1961–2000

Source: Annual reports of the Contrôle officiel suisse des chronomètres (COSC), 1961–2000.

Piaget watches. A boutique (1959), and then a manufacture (1960) were opened at Geneva, allowing the family firm better access to the luxury industry networks. However, the growth of Piaget was also enhanced by the liberalization of the industry which facilitated mergers. In 1964, Piaget bought up the company Baume & Mercier, another family firm from the Jura Mountains established at Geneva in 1920, and constituted the core of a luxury watches group which was be taken over by Cartier at the end of the 1980s.

Patek Philippe shows another path of development for the luxury watch industry, as at the beginning of the 21st century it is still a family firm.[286] Founded at Geneva in 1839 by a Polish refugee, Antoni Norbert Patek, who formed an alliance with the French engineer Jean Adrien Philippe in 1845, it was a small firm until World War II. During the years 1839–1900, its average output was a little bit below 1,800 watches a year. It grew at the beginning of the 20th century, reaching 6,050 watches in 1907 producing top-of-the-range goods, characterized by complicated watches and jewellery. Struggling with financial difficulties during the crises, in 1932 Patek Philippe was bought by the Stern family who still own it. The company was reorganized and experienced high growth during the second part of the 20th century, with output going up to 12,000 watches in 1970 and 20,000 in 2000. This relatively low level of production, together with stagnating productivity, is a good reflection of a luxury watch company's business model. Yet, the success of Patek Philippe also relies on a strategy close to that adopted by Rolex, albeit on a reduced scale. In 1932, Stern relaunched the company with the marketing of a new classic and basic model of watch, the Calavatra, commercialized in different varieties until today, and obviously standardized to be produced by machines. The union of excellence of manufacturing maintained by the creation of highly complicated watches, on which the brand strategy of the company is based, and the standardized production of a basic model, is at the root of Patek Philippe's success on the world market.

Table 22: Production and employees of Patek Philippe, 1920–2000

	1920	1932	1950	1970	1980	1990	2000
Employees (A)	–	–	164	366	298	324	650
Production (B)	2,500	683	6,650	12,000	10,000	15,000	20,000
Productivity (B/A)	–	–	40.5	32.8	33.6	46.3	30.8

Source: Statistics communicated by Patek Philippe.

4.5 The globalization of ownership and manufacturing

Decartelization made it possible for the Swiss watch industry to reorganize itself across national borders, and the second part of the 1960s was marked by a real internationalization of watch production. Actually, international expansion in watchmaking was not a new phenomenon. In particular, American companies, such as Bulova and Timex, had been organized on a global scale since the interwar period. As for Japanese firms, like Seiko or Citizen, they did the same from the beginning of the 1960s. However, the Swiss watch industry had been prevented from manufacturing outside Switzerland by the cartel legislation, which did not allow supplying parts from elsewhere than Switzerland before the middle of the 1960s.

The globalization of the Swiss watch industry was an answer to two important needs. First, watchmakers were looking to decrease production costs through the delocalization of a part of the production, essentially in South-East Asia. The manufacture of some parts, such as cases and metal bracelets, was largely moved abroad. Thus, in 1977, the Swiss production of watch metal bracelets amounted to only 228,000 pieces. In the same year, imports of this product reached 6.1 million pieces, mainly from Hong Kong (34%), Taiwan (23%) and South Korea (15%).[287] This transfer of production to Asia took a more important step after the foundation of the Swatch Group (1983). Reducing costs was a key issue in the competition against Japanese watch makers, and the company ETA SA (the new name of Ebauches SA, which was the producer of movements and parts for the Swatch Group, and the provider for nearly all the Swiss watch industry) opened a plant in Thailand as early as 1985. Other production facilities were then opened in Malaysia (1991) and in China (1996). In 1998, 33% of the employees of the Swatch Group were working in Asia.[288]

Second, we must consider strategic concerns. The Swiss acquisitions of interests in foreign watchmaking companies were used to access new outlets. It was, for example the strategy followed by ASUAG when it acquired interests in some French and German ébauche makers in the second part of the 1960s, with the aim of reinforcing the sale of its own products to these two neighboring countries. In France, ASUAG took over the watchmaker Lip.[289] In 1967–1968, it bought up both of the main French ébauches makers, the company France-Ébauches SA (45% of the French market in 1967), and the company Lip, with the objective of having access to new customers and benefiting from the distribution network of Lip in France, for the watches of its new group, GWC. However ASUAG did not invest in

Lip and the French watchmaker gradually lost its competitiveness. In 1973 its factory was occupied by the workers who ran the company for few months, and then the company was taken over by new investors before closing down in 1977. The investment by the group SGT into the American firms Waltham and Elgin at the beginning of the 1970s had a similar goal, namely to use their sale networks in the American market.

The internationalization of the Swiss watch industry was, however, not a unidirectional movement, only favorable to Swiss firms. As early as the second part of the 1960s, some foreign capital was invested in Switzerland. It was mainly the American companies who invested, with the takeover of Huguenin at Biel (1956) and of Büren Watch by Hamilton (1966), and then of Universal Geneve by Bulova (1967), and the acquisition of the group Movado-Zénith-Mondia by the American electronics company Zenith, which had no link with the Swiss watchmaker of the same name until then.[290] As for the Japanese watchmaker Seiko, it took over a small firm in Geneva, Jean Lassale (1985), with the ambition of launching into luxury watch production. The diversification into watchmaking of the foreign luxury groups Richemont and LVMH was also set in this framework. Finally, the intervention of some Chinese subcontractors of Hong Kong must be underlined, even if uncommon. They were not only mass producing cheap watch cases and metal bracelets but also developing a real industrial strategy, like the company Stelux Manufacturing Ltd, founded at Hong Kong in 1963 by a Chinese businessman.[291] In the 1960s, it manufactured dials, cases and bracelets for watchmakers around the world. The profits from these activities were invested in the middle of the 1970s in the takeover of various firms. Thus in Switzerland, this group controlled Unilux Holding SA, at Biel, which owned the companies Metalem SA (dials), Jean Vallon SA (cases), Orac SA (cases) Unilux SA (trade, service and design) and Stelux SA (assembly and sale of electronic watches). In 1976 it also invested in Bulova Watch, in the United States, and held 27% of its shares. Stelux Holding was in 2000 one of the main watch groups involved in the distribution of watches in Asia. It also owned three companies in Switzerland (Universal SA, Pronto Watch and Solvil & Titus) and some assembly plants in China and Hong Kong.[292]

Some subcontractors coping with globalization: the case makers

The delocalization of production had a direct influence on Swiss subcontractors, who were then facing serious competition. It was especially the case of the manufacturers of parts which necessitate only a simple production

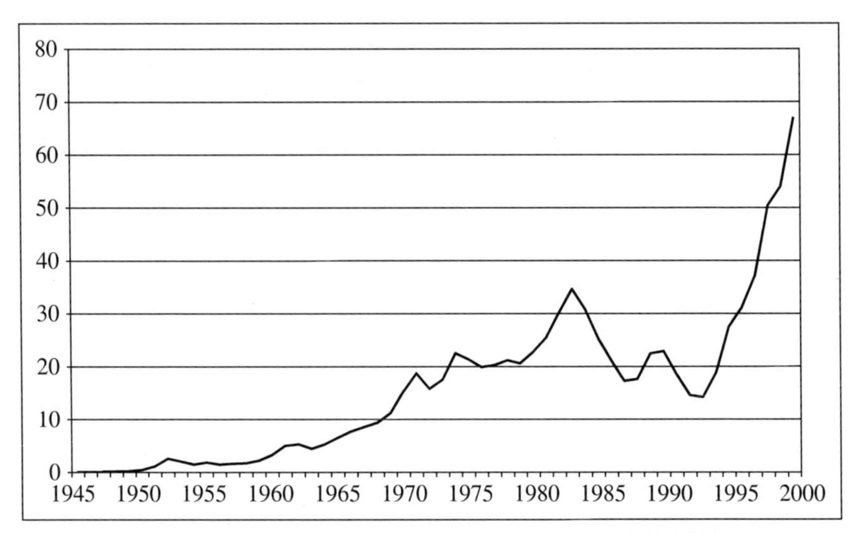

Figure 17: Swiss watches equipped with foreign cases, as a %, 1945–2000

Source: *Statistique du commerce de la Suisse avec l'étranger*, Berne : Département fédéral des Douanes, 1945–2000.

technology, like watch cases. Before the 1960s, the import of foreign cases was not completely forbidden by the cartel but it was mainly restricted to products manufactured by German firms who had had business relations with the Swiss watch industry since the late 19th century. In 1950, Germany supplied 87.4% of the all the imported watch cases.[293] However, these imports represented a negligible share of the market. In the 1950s, only 1.7% of all the Swiss watches exported were equipped with a foreign case. Thus, the cartel made it possible to have a very effective control of the market for case makers.

Table 23: Imports of watch cases in Switzerland, 1961–2000

	1961	1970	1980	1990	2000
Number of cases	1,575,405	8,040,596	6,485,811	6,951,382	20,061,662
Germany, as a %	47.2	15.2	8.4	2.5	0.2
France, as a %	24.8	12.7	26.3	15.9	1.5
Italy, as a %	3.3	3.4	8.0	5.2	45.2
Hong Kong, as a %	21.9	60.8	37.4	42.3	29.7
Thailand, as a %	0.0	0.0	8.2	16.3	3.8
China, as a %	0.0	0.0	0.0	0.0	17.6

Source: *Statistique du commerce de la Suisse avec l'étranger*, Berne : Département fédéral des Douanes, 1961–2000

The liberalization of the watch industry had a rapid impact on the Swiss market for watch cases. The number of imported cases grew from 83,000 pieces in 1950 to 1.6 million pieces in 1961, and then the growth accelerated and peaked in 1974 at 13.4 million pieces. The share of Swiss watches equipped with foreign cases went from 6.5% in 1965 up to 15.3% in 1970 and 21.4% in 1975. Thus an important part of the market was lost to domestic makers. Moreover, the dominant trend of this opening to foreign cases occurred within the context of a new globalization of production (see Table 23). The traditional suppliers of cases from Germany and France saw their position collapsing in the 1960s, a decade which established Hong Kong as a major production center, not only for Swiss watchmakers but also for their Japanese competitors.[294] Yet some Swiss industrialists were also engaged in this first wave of internationalization, in order to use cheap labor and be able to compete with foreign makers. At first, the Swiss watch case makers felt betrayed by this move. At the end of 1966, an agreement was signed between the Fédération Horlogère (FH) and entrepreneurs from Hong Kong. It notably planned technical assistance of the FH to Asian watch case makers, which led to the anger of the Union of Swiss watch case makers (Union des Fabricants Suisses de Boîtes, UFSB), a member of UBAH, which declared that *"the FH made fun of us. [It] gave what did not belong to it. It practically formalized in the bargain the free import of cases from Hong Kong to Switzerland."*[295] The development of case production in Hong Kong was, however, not only based on technical assistance given to Chinese entrepreneurs, but also on new firms founded at Hong Kong by Swiss watch groups. For example, in 1971 SSIH bought up the company Swiss Time Hong Kong, created two years before by industrialists from Basel.[296] In 1978, SSIH pursued this strategy in opening, together with Japanese industrialists, another case factory at Singapore, Precision Watchcase Ltd.[297] Moreover, in 1972 the Swiss government lowered the import tariff on products manufactured in Hong Kong and Singapore by 30%, thus favoring the import of watch parts.[298] Once again, in 1978 the UFSB denounced *"the united and doctrinaire free-traders [who] continue to successfully lobby the Federal division of commerce"*[299] but this reaction appeared as the last blaze of glory of an association whose political influence had vanished.

To try to counter to these Asian companies, some Swiss case makers established production plants in South-East Asia, with the aim of competing with these rivals. The creation in 1966 of the company Centre-Boîtes SA by a conglomerate of 39 Swiss watch case makers made it possible to open in 1968 a Swiss Watch Case Center in Hong Kong, to which was subcontracted the finishing of cases for the shareholders of the company.[300] Moreover, some individual makers set up production centers in Asia. A good example is the family Bourquard, owner of one of the biggest watch case enterprises of Switzerland, La Générale Holding SA, which had some ten plants and in 1969 had a production capacity of about 7 million pieces (that is, 14.5% of the Swiss production). After having imported, in the years 1967–1968, several tens of thousands of cases from a Chinese producer based at Singapore, Leung Lung Kee Metalware Ltd,[301] Bourquard set up, with him and an American customer, a watch case factory in Singapore, the Swiss Asiatic Co (1968), and then its own enterprise, Swiss Associated Industries Ltd (1969).[302] Other Swiss makers adopted the same strategy in the 1970s, as did for example Henri Paratte & Cie, supplier to the Japanese watchmaker Citizen, with the company Parathai at Bangkok (1972), and Ruedin SA, which took a share in Swisstime Philippines Inc. (1978)[303].

After the years of crisis and restructuring of the Swiss watch industry, during which a stagnation of case imports but an increase of their relative importance occurred (one third of Swiss watches were equipped with foreign cases in 1983), a second phase of globalization began, characterized by the nearly complete transfer of low- and middle-of-the-range production abroad. While the number of imported cases grew from 6.9 million pieces in 1990 to 20.1 million in 2000, the share of watches equipped with foreign cases experienced a real explosion during the 1990s, so that in 2000 two thirds of Swiss watches had a foreign case. This globalization of production relied mainly on production units based in Asia but one should also mention the growing importance of Italy in the 1990s, due to the company Lascor, which was bought up in 1996 by the Swatch Group and has supplied it since then with various cases (gold, silver and steel).

4.6 Towards luxury

Even if it was a low end watch – the Swatch – which made it possible for the Swiss watch industry to recover thanks to the adoption of a new marketing strategy, the evolution of the industry since the end of the 1980s is definitively characterized by a repositioning towards luxury. The foreign trade statistics emphasize this trend: the value was almost constantly growing

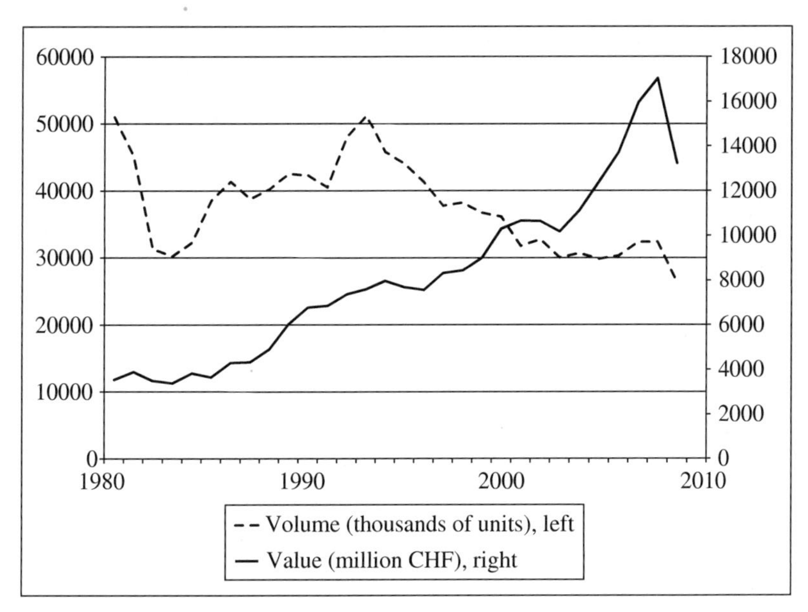

Figure 18: Swiss watch exports, value and volume, 1980–2009

Source: *Statistique du commerce de la Suisse avec l'étranger*, Berne : Département fédéral des Douanes, 1980–2009.

from 1987, while the volume has declined since 1994. The value of exports even increased at record rates at the beginning of the 21st century. It went from 3.7 billion francs in 1985 to 7.7 billion in 1995, 12.4 billion in 2005 and peaked at 17.0 billion in 2008 before coming back to 13.2 billion in 2009 due to the world crisis – the level of 2006. As for the number of exported watches, which in 1983 had reached its lowest level since 1950, it grew again in 1984, thanks to the Swatch, and peaked at 51.2 million pieces in 1993 – the level of 1965 – before entering a continuous declining trend. The volume of exports of watches was only 29.9 million pieces in 2005 and 26.4 million in 2009.

The evolution towards luxury appears as a consequence of the reorganization of the world watch market following the quartz revolution. The lowest end of the market was abandoned to newcomers, the producers of electronic watches from Hong Kong, China and India, who entered the market with extremely cheap products with which the Swiss makers could not compete. Nevertheless, the Swiss watch industry adopted a strategy of repositioning in the top of the market, concentrating its production on mechanical watches. The watch was no longer just a useful product or a luxurious jewel, it became a symbol of a tradition and of a technical culture.

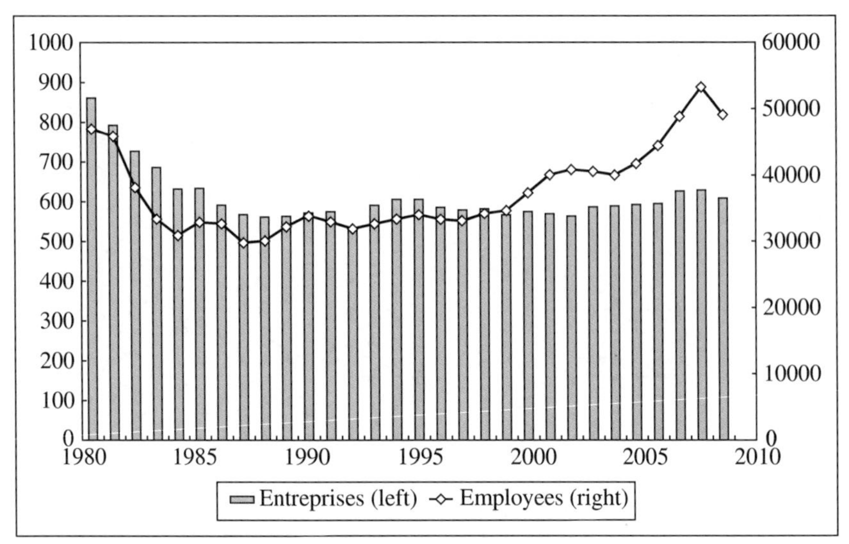

Figure 19: Enterprises and employment in the Swiss watch industry, 1980–2009

Source: Convention patronale, *Recensement 2009*, La Chaux-de-Fonds : CPIH, 2010, p. 9.

The complete watch companies were not the only firms affected by this repositioning towards luxury. It also happened in subcontracting companies which had not relocated their production in Asia. As the example of case makers shows, Swiss production was reoriented to gold cases. The proportion of enterprises active in gold cases went, for all the country, from 42.4% in 1975 to 50.0% in 1990 and 73.0% in 2000.

After a period marked by a large drop in employment and of the number of firms, at the end of the 1980s the industry entered a new phase of stability, with respect to the number of enterprises. Restructuring was over. There was an average number of 587 companies in the years 1990–2009. The new growth of the industry occurred within this new industrial framework. Employment experienced a new rapid growth trend, after having reached the minimum of 29,809 active workers in 1987. There were 33,923 active workers in 1990, 37,334 in 2000 and 41,478 in 2005. The average number of employees by firm was also growing: it exceeded 60 persons in 1999, recovering to the record level of the high growth period (1973–1974), and then 70 persons in 2005. In the years 2005–2009, the high growth of the Swiss watch industry led to a peak of 53,300 workers in 2008. Despite the recession due to the crisis, the number of workers was still 49,097 in 2009, that is, the level of 1979. Moreover, the average

Table 24: Outlets of the Swiss watch industry, 1960–2010, value as a %

<table>
<tr><th></th><th>1960</th><th>2000</th><th>2010</th></tr>
<tr><td>Europe</td><td>32.4</td><td>37.7</td><td>31.0</td></tr>
<tr><td>North America</td><td rowspan="2">40.2</td><td>18.6</td><td rowspan="2">14.0</td></tr>
<tr><td>Latin America</td><td>3.9</td></tr>
<tr><td>Asia-Oceania</td><td rowspan="2">22.4</td><td>33.8</td><td rowspan="2">54.0</td></tr>
<tr><td>Middle East</td><td>5.3</td></tr>
<tr><td>Africa</td><td>5.0</td><td>0.7</td><td>1.0</td></tr>
</table>

Source: *Statistique du commerce de la Suisse avec l'étranger*, Berne : Département fédéral des Douanes, 1960–2000.

number of employees per enterprise grew up steadily, reaching 85 in 2008 and 81 in 2009. The degree of concentration in the Swiss watch industry at the end of the 2000s is about twice what it was in the middle of the 1960s.

Finally, the reorientation towards luxury was accompanied by a strengthening of outlet diversification and, since the 2000s, a shift towards Asia, mainly China. North America (USA and Canada) lost its relative importance to Europe, and especially to Asia and Middle East. The liberalization of the Japanese watch market (1961), the economic growth of Middle-Eastern markets fuelled by oil-money, as well as the opening of China, are the main elements which explain the growing importance of the Orient for the export of Swiss watches. However, despite this shift, the Swiss watch industry was still very dependent on its traditional markets: thus, in 2009, the three main outlets were Hong Kong (16.4%), the United States (11.1%) and France (7.3%).[304] Yet these three markets amounted only to 34.8% of the value of the Swiss watch exports, while the three biggest outlets' share was 39.0% in 2001 (USA, Hong Kong, Japan) and 41.0% in 2005 (USA, Hong Kong, Japan). The growth at the end of the 2000s is thus accompanied by a diversification of markets, especially towards emerging markets. Since the middle of the decade, the share of China has become a key issue: it grew – Hong Kong not included – from a mere 0.3% of the total value of Swiss watch exports in 2001 to 2.6% in 2005 and 5.3% in 2009. This new Eldorado especially supports the development of big groups which have good access to the Chinese distribution system. In 2009, the net sales of the Swatch Group in the Greater China region (China, Hong Kong and Taiwan) amounted to 1.4 billion CHF, that is, 28% of the total gross sale of this company.[305] Besides, the emergence of Asia, and especially of China, as the most important outlet has supported the mutation of the Swiss watch industry towards luxury since the early 2000's.

Notes

237 *Statistique du commerce de la Suisse avec l'étranger*, Berne : Département fédéral des Douanes, 1945–1972.

238 Hampel, Heinz, *Automatic Armbanduhren aus der Schweiz: Uhren, die sich selbst aufziehen*, Munich: Callwey, 1992.

239 Donzé, Pierre-Yves, "The hybrid production system and the birth of the Japanese specialized industry: Watch production at Hattori & Co. (1900–1960)", *Enterprise & Society*, vol. 12 no. 2, 2011, p. 377.

240 *Statistique du commerce de la Suisse avec l'étranger*, Berne : Département fédéral des Douanes, 1974–1985.

241 Vogt, German, "Baumgartner, Arnold", *DHS*.

242 *Feuille fédérale*, 1970, pp. 722–723.

243 Archives cantonales jurassiennes (ACJ), Péquignot Papers, 122. Speach of Karl Huber, general secretary of the Federal Department of Public Economy, at a meeting with representatives of the watch industry organizations, 8 May 1959.

244 Perrenoud, Marc, "Gérard Bauer", *DHS*, <www.dhs.ch> (site accessed 26 May 2011).

245 *Feuille fédérale*, 1960, p. 1526.

246 Max Schmidheiny (1908–1991): member of the National Council (federal parliament), from St.Gallen, CEO of the cement companies Portland and Holderbank, member of the Board of Directors of Brown Boveri Co. Müller, Peter, "Schmidheiny, Max", *DHS*, <www.dhs.ch> (site accessed 26 June 2009).

247 *Bulletin sténographique des séances de l'Assemblée fédérale. Conseil national*, 13 juin 1961, pp. 202–203.

248 Schröter, Harm, "Cartels", *DHS*, <www.dhs.ch> (site accessed 15 June 2009).

249 *Feuille fédérale*, 1984, p. 848.

250 *Feuille fédérale*, 1961, p. 1604.

251 *Feuille fédérale*, 1980, p. 1329.

252 Stephens, Carlene and Dennis, Maggie, "Engineering time: inventing the electronic wristwatch", *British Journal for the History of Science*, vol. 33, 2000, pp. 477–497.

253 Knickerbocker, F., *Note on the watch industries in Switzerland, Japan, and the United States*, Harvard Business School, working paper, 1974, p. 10.

254 Perret, Thomas e.a., *Microtechniques et mutations horlogères : clairvoyance et ténacité dans l'Arc jurassien : un siècle de recherche communautaire à Neuchâtel*, Hauterive : G. Attinger, 2000, 333 p.

255 Pasquier, Hélène, *La "Recherche et Développement" en horlogerie. Acteurs, stratégies et choix technologiques dans l'Arc jurassien suisse (1900–1970)*, University of Neuchâtel, PhD thesis, 2007, p. 329.

256 Bauer Gérard e.a., *L'aventure de la montre à quartz : mutation technologique initiée par le Centre Electronique Horloger, Neuchâtel*, Neuchâtel : Centredoc, 2002, pp. 360–363.

257 *Hong Kong's Manufacturing Industries*, Hong Kong: Government Industry Department, 1994, p. 105.

258 Statistics based on the evaluations of the Seiko Institute of Horology, Tokyo.

259 Pasquier, Hélène, *La "Recherche et Développement" en horlogerie. Acteurs, stratégies et choix technologiques dans l'Arc jurassien suisse (1900–1970)*, University of Neuchâtel, PhD thesis, 2007.
260 *Seiko tokei no sengoshi*, Tokyo: Seiko, 1996, p. 154.
261 Statistics communicated by the Seiko Institute of Horology, Tokyo.
262 Donzé, Pierre-Yves, "The hybrid production system and the birth of the Japanese specialized industry: Watch production at Hattori & Co. (1900–1960)", Enterprise & Society, vol. 12 no. 2, 2011, pp. 356–397.
263 Statistics communicated by the Seiko Institute of Horology, Tokyo.
264 Hong Kong Census and Statistics Department, *Hong Kong Trade Statistics. Imports*, 1960–1970.
265 U.S. Department of Commerce, Bureau of the Census, *U.S. Imports commodity by country*, 1974–1976.
266 Kinugawa Megumu, *Nihon no baburu*, Tokyo: Nihon keizai hyoron, 2002.
267 Knickerbocker, F., *Note on the watch industries in Switzerland, Japan, and the United States*, Harvard Business School, working paper, 1974, p. 8.
268 Pasquier, Hélène, *La "Recherche et Développement" en horlogerie. Acteurs, stratégies et choix technologiques dans l'Arc jurassien suisse (1900–1970)*, University of Neuchâtel, PhD thesis, 2007, p. 44.
269 *La Suisse horlogère et revue internationale d'horlogerie*, décembre 1974–janvier 1975, pp. 39–40.
270 "Les concentrations dans l'industrie horlogère suisse", *Annales biennoises*, 1971, p. 50.
271 Donzé, Pierre-Yves, "Des montres et des pétrodollars : la politique commerciale d'une PME horlogère suisse. Aubry Frères SA, 1917–1993", *Revue suisse d'histoire*, 2004, pp. 384–409.
272 Donzé, Pierre-Yves, *The Comeback of the Swiss Watch Industry on the World Market: A Business History of the Swatch Group (1983–2010)*, Discussion Paper in Business and Economics, Osaka University, April 2011.
273 Japanese External Trade Organization (JETRO), Osaka, documentation on the Swiss watch industry.
274 Annual reports of the ASUAG, 1970–1979.
275 Fallet, Estelle, *Tissot, 150 ans d'histoire*, Le Locle : Tissot SA, 2003, p. 185.
276 Pasquier, Hélène, "Swatch Group", *DHS*, <www.dhs.ch> (site consulté le 25 juin 2009).
277 Fallet, Estelle, "Nicolas Hayek", *DHS*, <www.dhs.ch> (site consulté le 25 juin 2009).
278 Annual reports of the Swatch Group, <www.swatchgroup.com> (site accessed 26 May 2011).
279 Annual reports of the Richemont Group, <www.richemont.com> (site accessed 26 May 2011).
280 <www.lvmh.fr> (site accessed 26 May 2011).
281 Convention patronale, *Recensement 2007*, La Chaux-de-Fonds : CPIH, 2008, p. 10.
282 "1878–1978. Centenaire de la Manufacture des montres Rolex SA, Bienne", *Neues Bieler Jahrbuch*, 1979, pp. 101–109.
283 Knickerbocker, F., *Note on the watch industries in Switzerland, Japan, and the United States*, Harvard Business School, working paper, 1974, p. 9 and *Le Temps*, 27 December 2008.

284 Cardinal, Catherine (dir.), *L'art de mesurer le temps : Abraham-Louis Breguet, 1747–1823*, La Chaux-de-Fonds : MIH, 1997.
285 Fallet, Estelle, "Piaget", *DHS*, <www.dhs.ch> (site accessed 23 June 2009).
286 De Senarclens, Jean, "Patek Philippe", *DHS*, <www.dhs.ch> (site accessed 23 June 2009).
287 Japanese External Trade Organization (JETRO), Osaka, documentation on the Swiss watch industry.
288 Donzé, Pierre-Yves, *Histoire du Swatch Group*, Neuchâtel : Alphil-Presses universitaires suisses, 2012, pp. 52–57.
289 De Virieux, F.-H., *Lip. 100'000 montres sans patron*, Paris, Calmann-Lévy, 1973, 292 p.
290 Knickerbocker, F., *Note on the watch industries in Switzerland, Japan, and the United States*, Harvard Business School, working paper, 1974, pp. 18–19.
291 Japanese External Trade Organization (JETRO), Osaka, documentation on the Swiss watch industry.
292 Annual report of Stelux Holdings, 1999–2000, <www.irasia.com/listco/hk/stelux/annual/index.htm> (site accessed 28 December 2008).
293 *Statistique du commerce de la Suisse avec l'étranger,* Berne : Département fédéral des Douanes, 1950.
294 Blanc, Jean-François Blanc, *Suisse-Hong Kong. Le défi horloger. Innovation tech nologique et division internationale du travail*, Lausanne : Ed. d'en bas, 1988, pp. 94–132.
295 Archives of the UFSB, annual report, 1966.
296 Blanc, Jean-François Blanc, *Suisse-Hong Kong. Le défi horloger. Innovation technologique et division internationale du travail*, Lausanne : Ed. d'en bas, 1988, pp. 144–145.
297 Blanc, Jean-François Blanc, *Suisse-Hong Kong. Le défi horloger. Innovation technologique et division internationale du travail*, Lausanne : Ed. d'en bas, 1988, p. 145.
298 Archives of the UFSB, annual report, 1972.
299 Archives of the UFSB, annual report, 1978.
300 Archives of the UFSB, undated document (1969). See also Blanc, Jean-François Blanc, *Suisse-Hong Kong. Le défi horloger. Innovation technologique et division internationale du travail,* Lausanne: Ed. d'en bas, 1988, p. 149.
301 Centre jurassien d'archives et de recherches économiques (CEJARE), Bourquard Papers, correspondance of Humbert Bourquard, 1967–1969.
302 CEJARE, Fonds Bourquard, Letter from Humbert Bourquard to Sheldon Parker, New York, 10 September 1969.
303 Kleisl, Jean-Daniel, *Le patronat de la boîte de montre dans la vallée de Délémont : l'exemple de E. Piquerez S.A. et de G. Ruedin S.A. à Bassecourt (1926–1982)*, Delémont : Alphil, 1999, pp. 176–177.
304 *Statistique du commerce de la Suisse avec l'étranger,* Berne : Département fédéral des Douanes, 2005.
305 Swatch Group, *Annual report,* 2009, p. 177.

Conclusion

From the middle of the 19th century, the Swiss watch industry faced two major crises which challenged its existence.

The first was industrialization. American rivals challenged Swiss watchmakers by mass producing watches with machines in their factories and made Swiss watches become uncompetitive because they were too expensive or poorer quality. The industrialization of production modes was the object of intense discussions in Switzerland, and was opposed by many, but succeeded in being established during the 1880s and 1890s. After its successful move towards mechanization, during the interwar period the Swiss watch industry adopted an industrial policy aimed at putting an end to *chablonnage* and limiting the risk of industrial transplantation abroad. While customs protectionism spread worldwide from the 1890s, the export of disassembled watches, mainly to the United States, Japan, Russia and Germany, grew steadily and contributed to the development of new watchmaking companies in these countries. In order to cope with the emergence of new rival nations, and to maintain employment in Switzerland, watchmaking organized, under the aegis of the State, a cartel which forbade *chablonnage* and controlled the activities of watchmaking companies (the *Statut horloger*), as well as a trust for the control of the production of parts and movements (ASUAG).

The second existential crisis the Swiss watch industry faced was that of industrial concentration. The emergence and the spread of new mass production systems and new technologies, such as quartz watches, led to the establishment on the world market of rival firms in the United States (Timex) and in Japan (Seiko and Citizen). To remain competitive, the Swiss watch industry moved to large industrial concentration and restructuring in the 1960s–1980s which was characterized by the creation of new watch groups (Swatch Group, Richemont and LVMH).

During both these crises, the Swiss watch industry adopted, sometimes with difficulty, controls on its own structure, sometimes very restrictive, in order to keep industrial activity in Switzerland. The existence of the cartel from the 1920s to the 1960s, and then the legal definition of the Swiss Made label, resulted in watchmaking companies staying on Swiss territory. Despite the presence of American firms such as Bulova or Benrus from the 1900s, and despite the creation of production units in South-East Asia in the 1960s and 1970s, Swiss watchmaking is paradoxically an industry which has not globalized.

This lack of globalization of the Swiss watch industry is a key feature of this sector. Indeed, the other main export industries of the country, like chemicals, food or machines, all faced, from the end of the 19th century, a growth based on an early and pro-active globalization. Some Swiss multinational enterprises, like Ciba, Nestlé and Sulzer, founded their growth on the internationalization of their activities, with the creation of subsidiaries abroad and cooperation with foreign firms. It is this openness of the Swiss economy which is said to have enabled Switzerland to become one of the richest countries in the world in the 20th century.

The watch industry is obviously an exception due to its geographical localization. Why didn't this industry follow the path of globalization? The common explanation is that of the existence of a particular know-how, a specific technical culture, in the Jura Mountains, which explains the necessity, even for watch groups present throughout the world, to maintain production in Switzerland. However, this explanation is insufficient when we remember that the Japanese firm Seiko experienced success at the prestigious chronometry tests at the Neuchâtel (1967) and Geneva (1968) Observatory, and that these days some Chinese makers present tourbillion watches at the Basel Fair.

Two reasons explain why the Swiss watch industry was – and still is – localized in Switzerland. First, there was a strong political will to preserve the structure of family capitalism in the watch industry, especially during the period of the cartel (1930s–1960s), in order to maintain the social order. The watch industry being the main, even the only, industrial activity in the Jura Mountains, it was seen as a necessity to keep an industrial structure made up of hundreds of small firms which ensured the existence of workshops in each village and town, and thus employment. Second, after the collapse of the cartel and the passing of the legislation on Swiss Made (1971), there was a marketing reason. Together with its own restructuring into groups and its move towards luxury, the Swiss watch industry experienced the general mutation of the luxury industry, characterized by its democratization and the triumph of brands. In the globalized luxury industry, the territorial localization appears to be a key feature because it guarantees the image of tradition and excellence which are at the basis of its exclusivism. The tradition of luxury is European in its essence, within which watchmaking is Swiss.

References

1. Archival Sources

Archives fédérales suisses (Swiss Federal Archives, AFS), Bern E 7004, Watch Industry

Archives de l'État de Berne (Bern Cantonal Archives, AEB), Bern Papers of the Watchmaking School of Saint-Imier

Archives de l'État de Neuchâtel (Neuchâtel Cantonal Archives, AEN), Neuchâtel Department of Industry, Control of factories

Archives cantonales jurassiennes (Jura Cantonal Archives, ACJ), Porrentruy Péquignot Papers

Municipal archives, La Chaux-de-Fonds Papers of the Watchmaking School

Municipal archives, Le Locle Papers of the Civic guard

Association patronale de l'horlogerie et de la microtechnique (APHM), Bienne Papers of the Association cantonale bernoise des fabricants d'horlogerie (ACBFH)

Centre jurassien d'archives et de recherches économiques (CEJARE), Saint-Imier Bourquard Papers

Japan External Trade Organization (JETRO), Osaka Documentation on the Swiss watch industry.

Mémoire d'Ici (MDI), Saint-Imier Papers of the Watchmaking School of Saint-Imier Reports of Pfister on the watch company Longines

Musée international d'horlogerie, La Chaux-de-Fonds (MIH) Papers of the Museum Papers of the Société intercantonale des industries du Jura/Chambre suisse d'horlogerie

Schweizerische Sozialarchiv (SSA), Zurich Papers of the FTMH

Schweizerische Wirtschafts Archiv (SWA), Bâle H 45, Syndicat des fabricants de boîtes en argent

Seiko Institute of Horology, Tokyo Statistics on the Japanese watch industry

Union suisse pour l'habillage de la montre (USH), Bienne Annual reports of the Union des fabricants suisses de boîtes (UFSB) Papers of the UBAH

2. Published Sources

Annual reports, ASUAG, 1970–1979.

Annual reports, Contrôle officiel suisse des chronomètres, 1961–2005.

Annual reports, LVMH Group, <www.lvmh.fr>, 2000–2010.

Annual reports, Stelux Holdings, 1999–2000, <www.irasia.com/listco/hk/stelux/annual/index.htm>, 2000–2005.

Annual reports, Swatch Group, <www.swatchgroup.com>, 2000–2010.

Annual reports, Richemont Group, <www.richemont.com>, 2000–2010.

Bulletin sténographique des séances de l'Assemblée fédérale. Conseil national, 1961.

Convention patronale, *Recensement 2007*, La Chaux-de-Fonds : CPIH, 2008.

Dictionnaire historique de la Suisse, <www.dhs.ch>.

Exposition internationale de Saint Louis (U.S.A) 1904. Section française. Rapport du Groupe 32, Paris : Comité français des expositions à l'étranger, 1906, pp. 15–16.

Feuille fédérale, Berne, 1920–1984.

Feuille officielle suisse du commerce (FOSC), 1885–2005.

Hong Kong Census and Statistics Department, *Hong Kong Trade Statistics. Imports*, 1960–1970.

Hong Kong's Manufacturing Industries, Hong Kong : Government Industry Department, 1994.

Swiss Watch Yearbook (annuaire Davoine), 1870–1900.

Statistique du commerce de la Suisse avec l'étranger, Berne : Federal Department of Customs, 1885–2005.

United States, Bureau of Census, *Foreign Commerce and Navigation of the United States*, Washington: U.S. Govt Print. Off., 1890.

United States, Bureau of the Census, *U.S. Imports commodity by country*, 1965–1980.

3. Books and articles

BABEL Antony, *Histoire corporative de l'horlogerie, de l'orfèvrerie, et des industries annexes*, Genève : A. Jullien, 1916.

BARRELET Jean-Marc, "De la noce au turbin. Famille et développement de l'horlogerie aux XVIIIe et XIXe siècles", *Musée neuchâtelois*, 1994, pp. 213–226.

BAUER Gérard e.a., *L'aventure de la montre à quartz : mutation technologique initiée par le Centre Electronique Horloger, Neuchâtel*, Neuchâtel : Centredoc, 2002.

BECK Renatus (dir.), *Voies multiples, but unique. Regard sur le syndicat FTMH 1970–2000*, Lausanne : Payot, 2004.

BÉGUELIN Sylvie, "Naissance et développement de la montre-bracelet : histoire d'une conquête (1880–1950)", *Chronometrophilia*, 37 (1994), pp. 33–43.

BELINGER KONQUI Marianne, "L'horlogerie à Genève", in CARDINAL Catherine *e.a.* (dir.), *L'homme et le temps en Suisse, 1291–1991*, La Chaux-de-Fonds : Institut l'homme et le temps, 1991, pp. 123–129.

BERTHOUD Robert, *Répertoire des brevets*, s.l. : s.d.

BLANC Jean-François, *Suisse-Hong Kong. Le défi horloger. Innovation technologique et division internationale du travail,* Lausanne : Ed. d'en bas, 1988.

BLANCHARD Philippe, *L'établissage. Étude historique d'un système de production horloger en Suisse (1750–1950)*, University of Neuchâtel, PhD thesis, 2009.

BOLLI Jean-Jacques, *L'aspect horloger des relations commerciales américano-suisses de 1929 à 1950*, La Chaux-de-Fonds : La Suisse horlogère, 1956.

BORER Harry, "1878–1978. Centenaire de la Manufacture des montres Rolex SA, Bienne", *Neues Bieler Jahrbuch*, 1979, pp. 101–109.

BUBLOZ Gustave, *La Chaux-de-Fonds, métropole de l'industrie horlogère suisse*, La Chaux-de-Fonds : Société des fabricants d'horlogerie de La Chaux-de-Fonds, 1912.

Catalogue officiel illustré et explicatif de l'Exposition nationale d'horlogerie et internationale de machines et outils employés en horlogerie en juillet 1881 à La Chaux-de-Fonds sous le patronage de la Société d'émulation industrielle, La Chaux-de-Fonds : Imp. du National suisse, 1881.

CHAPUIS Alfred, *La montre chinoise*, Neuchâtel : Attinger Frères, 1919.

CHOU Kouken, *Nihon tokeisangyo no hatten to kigyo katsudo*, University of Kyoto, MA thesis, 2002.

Cinquantenaire de l'École d'horlogerie de Saint-Imier, Saint-Imier, 1916.

CUTMORE M., *Watches 1850–1980*, London: David & Charles, 1989

Daumas Jean-Claude, "Districts industriels : le concept et l'histoire", Helsinki, XIVᵉ World Economic History Congress, 2006, <http://www.helsinki.fi/iehc2006/papers1/Daumas28.pdf>.

Davies Alun C., "British Watchmaking and the American System", *Business History*, 1993, n° 35/1, pp. 40–54.

De virieux F.-H., *Lip. 100 000 montres sans patron*, Paris : Calmann-Lévy, 1973.

Donzé Pierre-Yves, "Des montres et des pétrodollars : la politique commerciale d'une PME horlogère suisse. Aubry Frères SA, 1917–1993", *Revue suisse d'histoire*, 2004, pp. 384–409.

Donzé Pierre-Yves, "Les industriels horlogers du Locle (1850–1920), un cas représentatif de la diversité du patronat de l'Arc jurassien", in Daumas Jean-Claude (dir.), *Les systèmes productifs dans l'Arc jurassien. Acteurs, pratiques et territoires (XIXᵉ–XXᵉ siècles),* Besançon, Presses universitaires de Franche-Comté, 2005, pp. 61–82.

Donzé Pierre-Yves, "Le Japon et l'industrie horlogère suisse. Un cas de transfert de technologie durant les années 1880–1940", *Histoire, Économie et Société*, 2006, pp. 105–125.

Donzé Pierre-Yves, *Les patrons horlogers de La Chaux-de-Fonds (1840–1920). Dynamique sociale d'une élite industrielle*, Neuchâtel : Alphil, 2007.

Donzé Pierre-Yves, "The hybrid production system and the birth of the Japanese specialized industry: Watch production at Hattori & Co. (1900–1960)", *Enterprise & Society*, vol. 12 no. 2, 2011, pp. 356–397.

Donzé Pierre-Yves, *A Business History of the Swatch Group: The Rebirth of Swiss Watchmaking and the Globalization of the Luxury Industry,* Basingstoke: Macmillan, 2014.

Donzé Pierre-Yves, *Histoire du Swatch Group,* Neuchâtel: Alphil-Presses universitaires suisses, 2012.

Dubois Gérard, *Les débuts du syndicalisme horloger dans Les Franches-Montagnes (1886–1915)*, University of Geneva, MA thesis, 1984.

Ezawa Tomikichi, *Nanajûnana ô kaikodan*, Tokyo : Shikaishobo, 1939.

Fallet Estelle, *Tissot, 150 ans d'histoire*, Le Locle : Tissot SA, 2003.

Fallet Estelle and Cortat Alain, *Apprendre l'horlogerie dans les Montagnes neuchâteloises, 1740–1810*, La Chaux-de-Fonds : Institut l'homme et le temps, 2001.

Fallet-Scheurer Marius, *Le travail à domicile dans l'horlogerie suisse et ses industries annexes*, Berne : Imp. de l'Union, 1912.

FRAGOMICHELAKIS Michel, *Culture technique et développement régional. Les savoir-faire dans l'Arc jurassien*, Neuchâtel : ISSP, 1994.

FRANCILLON André, *Histoire de la fabrique des Longines, précédée d'un essai sur le comptoir Agassiz*, Saint-Imier : Longines, 1947.

GERBER Jean-Frédéric, "Le syndicalisme ouvrier dans l'industrie suisse de la montre de 1880 à 1915", in GRUNER Erich, *Arbeiterschaft und Wirtschaft in der Schweiz, 1880–1914*, Zurich : Chronos, 1988, vol. 2, pp. 479–528.

GUEX Sébastien, "À propos des gardes civiques et de leur financement à l'issue de la Première Guerre mondiale", in *Pour une histoire des gens sans histoire : ouvriers, exclues et rebelles en Suisse, XIXe–XXe siècles*, Lausanne : Editions d'en bas, 1995, pp. 255–264.

GAGNEBIN-DIACON Christine, *La fabrique et le village : la Tavannes Watch Co, 1890–1918*, Porrentruy : CEH, 2006 (2nd ed.).

HAMPEL, Heinz, *Automatic Armbanduhren aus der Schweiz: Uhren, die sich selbst aufziehen*, Munich: Callwey, 1992.

HARROLD Michael C., *American Watchmaking. A Technical History of the American Watch Industry, 1850–1930*, Columbia: NAWCC, 1984.

HENRY BÉDAT Jacqueline, *Une région, une passion : l'horlogerie. Une entreprise : Longines*, Saint-Imier : Longines, 1992.

HOKE Donald R., *Ingenious Yankees. The Rise of the American System of Manufacturers in the Private Sector*, New York: Columbia University Press, 1990.

HOSTETTLER Patricia, *Naissance et croissance d'une manufacture horlogère : la fabrique de montres Zénith SA au Locle (1865–1925)*, University of Neuchâtel, MA thesis, 1987.

HUMAIR Cédric, *Développement économique et État central (1815–1914) : un siècle de politique douanière suisse au service des élites*, Berne : Lang, 2004.

HUMAIR Cédric, *1848 : Naissance de la Suisse moderne*, Lausanne : Antipodes, 2009.

JAQUET Eugène, *L'école d'horlogerie de Genève, 1824–1924*, Genève : Editions Atar, 1924.

JEQUIER François, *De la forge à la manufacture horlogère (XVIIIe–XXe siècles). Cinq générations d'entrepreneurs de la vallée de Joux au cœur d'une mutation industrielle*, Lausanne : Bibliothèque historique vaudoise, 1983.

JOSEPH Roger, "La naissance de la paix du travail", in *L'homme et le temps en Suisse, 1291–1991*, La Chaux-de-Fonds : IHT, 1991, pp. 259–264.

KLEISL Jean-Daniel, *Le patronat de la boîte de montre dans la vallée de Délémont : l'exemple de E. Piquerez S.A. et de G. Ruedin S.A. à Bassecourt (1926–1982),* Delémont : Alphil, 1999.

KNICKERBOCKER F., *Note on the watch industries in Switzerland, Japan, and the United States*, Harvard Business School, working paper, 1974.

KOHLER François, "La grève générale dans le Jura", in VUILLEUMIER Marc (dir.), *La grève générale de 1918 en Suisse*, Genève : Grounauer, 1977, pp. 61–78.

KOHLER François, "Les communautés juives dans le Jura (XIXe–XXe siècles)", in *L'Hôtâ*, n° 20, 1996, pp. 73–84.

KNOBEL Joëlle, *Une manufacture d'horlogerie biennoise : la Société Louis Brandt & Frère (Omega), 1895–1935*, University of Neuchâtel, MA thesis, 1997.

KOLLER Christophe, *"De la lime à la machine". L'industrialisation et l'État au pays de l'horlogerie. Contribution à l'histoire économique et sociale d'une région suisse*, Courrendlin : CSE, 2003.

KURZ O., *European Clocks and Watches in the Near East,* London: Warburg Institute, 1975.

LAMARD Pierre, *Histoire d'un capital familial au XIXe siècle : le capital Japy (1777–1910),* Belfort : Société belfortaine d'émulation, 1988.

LANDES David S., *L'heure qu'il est : les horloges, la mesure du temps et la formation du monde moderne*, Paris : Gallimard, 1988.

"Les concentrations dans l'industrie horlogère suisse", *Annales biennoises*, 1971.

Les écoles suisses d'horlogerie, Zurich : Fritz Lindner, 1948.

Les Ébauches : Festschrift zur Feier des fünfundzwanzigjährigen Bestehens der Ebauches-A.G., s.l., s.n., 1951.

LINDER Patrick, *Organisation et technologie : un système industriel en mutation. L'horlogerie à Saint-Imier, 1865–1918*, University of Neuchâtel, MA thesis, 2006.

LINDER Patrick, *Au cœur d'une vocation industrielle : les mouvements de montre de la maison Longines : (1832–2007) : tradition, savoir-faire, innovation,* Saint-Imier : Édition des Longines, 2007.

L'Observatoire cantonal neuchâtelois, 1858–1912, Neuchâtel : DIP, 1912.

LUCIRI Pierre, "L'industrie suisse à la rescousse des armées alliées. Un épisode de la coopération interalliée pendant l'été 1915", *Relations internationales*, 1974, pp. 99–114.

LÜTHY Herbert, *La Banque protestante en France de la Révocation de l'Édit de Nantes à la Révolution*, 2 volumes, Paris : SEVPEN, 1959–1961.

MAEDER Alain, *Gouvernantes et précepteur neuchâtelois dans l'empire russe (1800–1890)*, Neuchâtel : Institut d'histoire, 1993.

MAKOTO Ichihara, *Yume o utta otoko. Kindai sangyo no paionia. Tenshodo – Ezawa Kingoro*, Tokyo : Ronshobo, 1990.

MARION, Gilbert, "Alfred Lugrin", *DHS*.

MARTI Laurence, "Nicolas Junker, Fabrique de machines, Moutier (1883–1905) ou les difficultés d'une entreprise innovante à la fin du XIXe siècle", in TISSOT Laurent (ed.), "Entreprises et réseaux. Les acteurs de l'industrialisation dans l'Arc jurassien", *Actes de la Société jurassienne d'émulation*, 1999, pp. 298–305.

MARTI Laurence, "Le tour à poupée mobile. (Jura suisse, 1870–1920)" in BELOT Robert, COTTE Michel and LAMARD Pierre (ed.), *La technologie au risque de l'histoire*, Belfort-Montbéliard : UTBM, 2000, pp. 191–198.

MARTI Laurence, *L'invention de l'horloger. De l'histoire au mythe de Daniel JeanRichard*, Lausanne : Antipodes, 2003.

MARTI Laurence, *Une région au rythme du temps. Histoire socio-économique du Vallon de Saint-Imier et environs, 1700–2007*, Saint-Imier : Édition des Longines, 2007.

NICOLET Georges, *Au cours du temps. Nivarox-FAR, 150 ans d'histoire*, Le Locle : Nivarox, 2000.

OZAKI Mayako, "18 seiki kohan junebu-shi no inyumin ni okeru shusshinchi sokugyo kosei no tenkan to renzoku", *Shakai-keizai-shi gaku*, vol. 71 no. 2, 2005.

PASQUIER Hélène, *La "Recherche et Développement" en horlogerie. Acteurs, stratégies et choix technologiques dans l'Arc jurassien suisse (1900–1970)*, University of Neuchâtel, PhD thesis, 2007.

PELLATON Jean, *Centenaire de la fabrication de l'assortiment à ancre au Locle, 1850–1950*, Le Locle : FAR, 1950.

PERRENOUD Marc, "Crises horlogères et interventions étatiques : le cas de la Banque cantonale neuchâteloise dans l'entre-deux-guerres", in CASSIS Youssef and TANNER Jakob (ed.), *Banken und Kredit in der Schweiz, 1850–1930*, Zurich : Chronos, 1993, pp. 209–240.

PERRENOUD Marc, "L'évolution industrielle de 1914 à nos jours", in *Histoire du Pays de Neuchâtel*, Hauterive : Gilles Attinger, vol. 3, 1993, pp. 146–155.

PERRET Thomas *e.a.*, *Microtechniques et mutations horlogères : clairvoyance et ténacité dans l'Arc jurassien : un siècle de recherche communautaire à Neuchâtel*, Hauterive : G. Attinger, 2000.

Petitpierre Alphonse, *Un demi-siècle de l'histoire économique de Neuchâtel, 1791–1848*, Neuchâtel : Librairie Jules Sandoz, 1871.

Pinot Robert, *Paysans et horlogers jurassiens*, Genève : Gronauer, 1979 (2nd ed.).

Rapport à la Société intercantonale des industries du Jura sur la fabrication de l'horlogerie aux États-Unis, 1876, Saint-Imier : Longines, 1992.

Richon Marco, *Omega Saga*, Bienne : Fondation Adrien Brandt, 1998.

Rieben Henri, Urech Madeleine and Iffland Charles, *L'horlogerie et l'Europe*, Neuchâtel : La Baconnière, 1959.

Ritzmann Heiner (ed.), *Statistique historique de la Suisse*, Zurich : Chronos, 1996.

Scheurer Frédéric, *Les crises de l'industrie horlogère dans le canton de Neuchâtel*, La Neuveville, Ed. Beerstecher, 1914.

Scheurer Hugues, "Paysans-horlogers : mythe ou réalité ?", in mayaud Jean-Luc and henry Philippe (ed.), *Horlogeries. Le temps de l'histoire*, Besançon : Annales littéraires de l'Université de Besançon, 1996, pp. 45–54.

Scranton Philippe, *Endless Novelty. Specialty Production and American Industrialization, 1865–1925*, Princeton: Princeton University Press, 1997.

Seiko tokei no sengoshi, Tokyo: Seiko, 1996.

Shashi, Tokyo: Citizen Watch Co, 2 vol., 2002.

Société générale de l'horlogerie suisse SA. ASUAG. Historique publié à l'occasion de son vingt-cinquième anniversaire, 1931–1956, Bienne : Arts graphiques SA, 1956.

Stephens Carlene and Dennis Maggie, "Engineering time: inventing the electronic wristwatch", *British Journal for the History of Science*, vol. 33, 2000, pp. 477–497.

Tissot Charles-Émile, *Rapport spécial sur l'exposition d'horlogerie*, s.l. : s.n., 1894.

Trueb Lucien F., *125 ans de chronométrage Longines : l'équité dans la mesure du temps, l'élégance dans le sport*, St-Imier : Edition des Longines, 200.

Uchida Hoshimi, *Tokei sangyou no hattatsu*, Tokyo: Seiko Institute, 1985.

Uttinger Hans W. and Papera D. Robert, "Threats on the Swiss Watch Cartel", *Western Economic Journal*, 1965, pp. 200–216.

Veyrassat Béatrice, *Réseaux d'affaires internationaux, émigrations et exportations en Amérique latine au XIXe siècle. Le commerce suisse aux Amériques*, Genève: Droz, 1994

VEYRASSAT Béatrice, "Manufacturing flexibility in nineteenth-century Switzerland: social and institutional foundations of decline and revival in calico-printing and watchmaking", in SABEL Charles F. and ZEITLIN Jonathan (ed.), *World of possibilities. Flexibility and Mass Production in Western Industrialization*, New York: Cambridge University Press, 1997, pp. 188–237.

VOGT, German, "Baumgartner, Arnold", *DHS*.

WATKINS Richard, *Watchmaking and the American System of Manufacturing*, Tasmania: Richard Watkins, 2009, <www.watkinsr.id.au/AmSystem.pdf>.

WYSS Jean, *La création de l'Union des Branches Annexes de l'Horlogerie (UBAH) et les vingt premières années de son activité (1927–1947),* La Chaux-de-Fonds : UBAH, 1947.